# INTERTWINED CREATURES

# INTERTWINED CREATURES

*The Embodied Cognitive Science*
*of Self and Other*

ANTHONY CHEMERO

Columbia University Press
*New York*

Columbia University Press
*Publishers Since 1893*
New York    Chichester, West Sussex
cup.columbia.edu

Cataloging-in-Publication Data available from the
Library of Congress.

ISBN 9780231195386 (hardback)
ISBN 9780231223195 (trade paperback)
ISBN 9780231538800 (epub)
ISBN 9780231565226 (PDF)

LCCN 2025028864

Cover design: Julia Kushnirsky
Cover image: Andrea Chemero, *Tree Stump Node*,
watercolor on paper, 2020.

GPSR Authorized Representative: Easy Access System
Europe, Mustamäe tee 50, 10621 Tallinn, Estonia,
gpsr.requests@easproject.com

*For Andrea*

# CONTENTS

*Preface*   *ix*

**PART I**

1   Other Minds   3

2   The Embodied Mind   18

3   The Intertwined Self   35

**PART II**

4   Radical Embodied Cognitive Science   55

5   Synergies and the Intertwined Self   66

**PART III**

6   Social Ontology, Representation Hunger,
    and the Intertwined Self   91

# CONTENTS

7  The Pragmatist Tradition and Inner Speech  111

8  Reorienting Ethics and Political Theory
   Around the Intertwined Self  127

9  Coda: Blanks Among Us  150

**APPENDIXES**

*Appendix for People Who Like Math*  157
*Meta-Appendix: The Controversy over 1/f Noise*  169

*Notes*  183
*Bibliography*  209
*Index*  231

# PREFACE

IN 2015, the great linguist, cognitive scientist, and political theorist Noam Chomsky published a book called *What Kind of Creatures Are We?*[1] For those of us who have been keeping up with Chomsky's work, it was unsurprising that he gave the same answer to this question that he would have given in the 1950s. The kind of creatures we are, according to Chomsky, is a matter of our internal language capacities, which evolved for thinking and whose activities rarely lead us to produce external speech. That is, we are thinkers, and most of our activities as thinkers are hidden from others. As Chomsky knows, this view is nearly identical to the view developed by philosophers of the Modern era (approximately 1550 to 1900), especially René Descartes. I call this view "the modern theory of the mind." The basic idea of the modern theory of the mind is that your mind is hidden from others and only accidentally related to your body, environment, and other people. This understanding of the mind is crucial in the founding of psychology as a science. It is also important in the founding of Western political theory. Chomsky is hardly the only person currently defending the modern view of the mind; he is joined in this by most cognitive scientists, neuroscientists, and economists.

I have never been satisfied with this view. The primary influence on my early thinking in philosophy was my teacher Daniel Dennett. Dennett never had patience for the idea that our minds are a product of an innate language capacity that evolved for thinking. Minds, he thought, are necessarily partly social. According to what he called the intentional stance, having a mind is a matter of being the sort of thing whose behavior can be explained by the attribution of mental states.[2] Part of Dennett's goal in understanding minds as being independent of the internal processes of our brain was ethical. Dennett saw our understanding of ourselves and others as having minds as indispensable to our moral and legal systems and was guarding against claims that neuroscientists do not find thoughts, desires, or emotions when they look into brains. Our having thoughts, desires, and emotions, according to the view I inherited from Dennett, does not depend on any findings from neuroscience. Later, I thought that the real problem with ideas like those of Descartes and Chomsky was their assumption that our ability to have hidden, inner thoughts is the result of an innate abilities. This sort of nativist thinking is morally and politically problematic, often leading its proponents to make straightforwardly racist and sexist claims.[3]

I will argue in this book that the modern theory of mind is incorrect and give a different answer to Chomsky's question.

We are intertwined creatures. To make the case for this, I will develop a conception of what the mind is that centers embodiment and social interaction. According to what I will call "the intertwined self," having a mind as an individual depends on your body and your interactions with the environment, tools, and other people. It is our history of engaging with other people that enables us to have minds on our own. I will offer some philosophical arguments and present some empirical research supporting the intertwined self. I will then look at some philosophical consequences of the intertwined self. I will also connect the view to works by phenomenological philosophers, pragmatists, so-called continental philosophers, and feminist political philosophers. My intent in all of this is to engage in a version of the ameliorative rethinking of race and gender categories by feminist philosopher Sally Haslanger[4] but for "self," "mind," and person," and with a lot more scientific data.

In chapter 1, I introduce the traditional philosophical problem of other minds using the 2013 film *The World's End* (directed by Edgar Wright).[5] As the problem of other minds goes, how do we know that other people have minds when we can see only their behaving bodies? This is an odd, even pathological thought that we each might be somehow cut off from the rest of humanity, yet its roots are very old. The modern theory of the mind as inner and invisible that enables this philosophical problem goes back to Descartes at least. After briefly introducing the modern theory of the mind and its import in the cognitive sciences and political theory, in chapter 2, I discuss a very different approach to our knowledge of other minds that begins with a rejection of the modern theory of the mind as inner and invisible: humans are always in the world with other minded people, and each human experiences those others as minded without having to theorize or simulate. This alternative view is based on the work of phenomenological philosophers and is currently defended by those who endorse ecological or enactive approaches in philosophy and the cognitive sciences. Proponents of these views argue that the object of a scientific psychology is not just a brain or even a biological body but a body-in-an-environment, where the environment includes technology and other humans. These two views are explicit rejections of the modern theory of the mind. In chapter 3 I argue that, according to the ecological and enactive approaches, the most basic possible understanding of a mind, what has been called the minimal self, is already embodied and social. This is an endorsement of what Maurice Merleau-Ponty, following Edmund Husserl, called the *Ineinander*: the self and world and the self and the other are in one another.

The next two chapters marshal scientific evidence for the intertwined self. In chapter 4, I introduce what I have called "radical embodied cognitive science," an approach to doing scientific psychology that preserves the insights of phenomenological philosophers. Radical embodied cognitive science combines ideas from ecological psychology, enactivism, and nonlinear dynamical systems theory. It explains the actions of humans interacting with their environments and other humans, without explaining them in terms of inner, invisible minds. In chapter 5, I introduce the concept of synergy, from nonlinear dynamical systems theory. The concept of "synergy" used here is drawn from twentieth-century advances in the physics of far-from-equilibrium, complex systems and is generally used as a synonym for "complex system." In a synergy, energetic and/or chemical constraints are applied to a system, causing some of its components to form units that work together. Cells are synergies; organs are synergies; organisms are synergies. Living things are generally hierarchies of synergies in which the synergy at one scale (e.g., an organ) is a temporary collection of synergies at a smaller scale (e.g., cells). I turn then to interpersonal synergies. I present evidence that humans skillfully and unreflectively form synergies with other humans across a wide variety of tasks. This is empirical evidence for the intertwined self.

The final chapters collect some consequences of the philosophical and empirical arguments in chapters 1–5. Chapter 6 begins with a recap. The fact that there are human-tool and interpersonal synergies implies that the boundaries of a human being are flexible. Sometimes what counts as a thinking thing or an agent is a brain in a body; sometimes, it is a body and a tool; sometimes, it is two or more humans acting as a unit. The fact that our being in the world with others is so fundamental suggests that humans are essentially embodied and social. The self is in the world and in others; the world and others are in the self. This view not only contradicts the modern understanding of the mind as inner and invisible; it also contradicts the conceptions of the self and the person that are based on them. It has consequences for philosophical controversies around group cognition, and it leads to different understandings of imagination and creativity. In chapter 7, I give an account of our private, inner speech that begins with the intertwined self. I draw connections between this view and those held by pragmatists like George Herbert Mead and Soviet psychologist Lev Vygotsky.

The fact that we can directly perceive aspects of the minds of others does not imply that none of our thoughts are private. Chapter 8 connects the intertwined self to politics and ethics. In particular, because much of Western political theory assumes a conception of humans derived from John Locke's theory of the self, which is an instance of the modern theory of the mind, the preceding chapters suggest a need for revision. Continental and feminist thinkers such as Gilles Deleuze, Donna Haraway, Lorraine Code, Moira Gatens, Carol Gilligan, and Annette Baier have already taken the embodied, social self into account in their political philosophies. The arguments of these thinkers are not intended as ends in themselves but rather as entry points into new ethical and political arguments. My arguments in the preceding chapters also support the feminist ethical and political positions that follow from them. Chapter 9 offers a brief coda.

There are two appendixes. The first contains the technical material suppressed in chapters 4 and 5. The second appendix addresses a controversy related to one of the data analyses described in the first appendix.

## ACKNOWLEDGMENTS

I have been working on this book for a long time, and a lot of people helped out. Mike Anderson, Louise Barrett, Abeba Birhane, Fred Cummins, Shaun Gallagher, Michael Silberstein, and Ashley Walton have been with me from the very beginning. Mike and Fred read several versions of the whole manuscript.

Ed Baggs disagrees with much of what is in this book but gave me valuable comments on the whole thing. I work closely with graduate students, many of whom were important in developing the ideas in this book, including Luis Favela, Maurice Lamb, Vicente Raja, Gui Sanches de Oliveira, Chris Riehm, Patrick Nalepka, Patric Nordbeck, Amanda Corris, Jay McKinney, Frank Faries, Johanna Webb, Sahar Heydari Fard, Emilien Dereclenne, Mark Órnelas, Daniel Mattox, Elmo Feiten, Zach Peck, Carlos Munoz, and Taraneh Wilkinson. Special thanks go to the Comrades.

The ideas in this book are better than they would have been without advice from Jay Holden, Mike Riley, Kevin Shockley, Paula Silva, Tehran Davis, Tamara Lorenz, Mike Richardson, Rachel Kallen, Heidi Kloos, Heidi Maibom, Tom Polger, Peter Langland-Hassan, Zvi Biener, Angela Potochnik, Kris Holland, Kate Sorrels, Alexandra Paxton, Chris Kello, Rick Dale, Drew Abney, Michael Spivey, Carolyn Dicey Jennings, Jeff Yoshimi, Paul Smaldino, John Sutton, Melina Gastelum, Sergio Martinez, Maria Baghramiam, Jim O'Shea, Erik Rietveld, Jelle Bruineberg, Julian Kiverstein, Miriam Kyselo, Ezequiel di Paolo, Somogy Varga, Deb Tollefsen, Chris Baber, Simon Penny, David Kirsh, Mog Stapleton, Mark James, Anna Ciaunica, Sanneke de Haan, Charles Lenay, Joel Krueger, Michele Maiese, Takashi Ikegami, Shogo Tanaka, Ivana Konvalinka, Tom Froese, Tetsuya Kono, Satoshi Sako, Sally Haslanger, Dan Smith, Javier Gomez-Lavin, J.P. Messina, Evan Westra, Sarah Robins, Corey Maley, Brett Karlan, Dan Kelly, Marta Caravà, John Protevi, Kathie Galotti, Julie Neiworth, Mija van der Wege, Jason Decker, Harry Heft, Joanna Rączaszek-Leonardi,

Alva Noë, Sune Steffensen, Peter Henzi, Erik Myin, Dan Hutto, Miguel Segundo-Ortin, Manuel Heras-Escribano, Lorena Lobo, and Shay Welch.

Wendy Lochner has been a wonderful (and patient) editor. Most editors at university presses just acquire books; Wendy actually edits and makes them better. Thanks to Evan Thompson and Adrian Parr for telling me how great it is to work with Wendy. Thanks to Marianne L'Abbate and Ben Kolstad for copyediting the final manuscript and to Taraneh Wilkinson for help with the page proofs and the index.

While I was writing this book, two of my most important mentors died: Dan Dennett and Mike Turvey. Their names appear many times in this book. I was lucky to have them as teachers and proud to have them as friends.

The writing of this book was supported by the Taft Center for the Humanities at the University of Cincinnati.

This, not just the book but everything, would have been impossible without my family: Andrea Chemero, Ava Chemero, Henry Chemero. Special thanks to Andrea for the cover art—and more.

# PART I

*There is a Zulu phrase,* 'Umuntu ngumuntu ngabantu', *which means "A person is a person through other persons." This is a richer and better account, I think, than "I think, therefore I am."*

Abeba Birhane, *Aeon* (2017)

*Chapter One*

## OTHER MINDS

**THIS CHAPTER** contrasts two broad views of the nature of the human mind, especially in how they relate to the philosophical problem of other minds.

### AVOIDING BLANKS

*The World's End*, the feature film that is the third installment of Edgar Wright's "Cornetto Trilogy," exemplifies a long-standing philosophical problem. In the movie, a group of not-very-lovable losers returns to their hometown of Newton Haven to try to complete the pub crawl that they had failed to finish twenty years earlier. In one key scene, Gary King, the least lovable among them, played by Simon Pegg, starts a fight with a gruff teen in a pub toilet. During the fight and in the subsequent brawl, it becomes clear that at least some, or maybe most or all, of the town's residents are humanoid robots, which the pub crawlers dub "blanks." For the rest of the movie, the pub crawlers are never certain whether the Newton Haven residents they meet are actually blanks instead of humans. The friends inevitably become suspicious of one another as well. (I won't give away more here, but I will say that it gets much stranger from this point onward.) This is inevitable because it seems you can never know, not really, what is going on inside someone else in the way that you can know what is happening inside yourself. You know your own experiences "from the inside": you can never have that kind of access to anyone else's experiences. This is the philosophical *problem of other minds*. I know about my own experiences by virtue of having them. I don't have anyone else's experiences "from the inside," so I don't know them in the same way. This difference is not hard to see. We all have been misunderstood from time to time, for example, when an intended joke is not taken as one. This happens because we know our own intentions in a way that others do not. Sometimes philosophers say that we have *privileged access* to our own, and only our own experiences.

This is taken by most philosophers to be a genuine problem. I have privileged access to my thoughts and only my thoughts, and although there are sometimes misunderstandings with other people, they are the exception. More often than not, we know what people with whom we interact are experiencing because they tell us, and if we want to know more, we can ask. And we know that they are like us—made of the same biological stuff, in the same geographic location, feeling the same weather, seeing and hearing and smelling similar things. When things happen to other people that are not happening to us, we can figure out what it is like for them. It hurts to get hit with a baseball, so that person is experiencing pain; you can tell that another person smells something bad by the face they are making. To be clear, in the 2020s, we do interact with blanks regularly, but they are artificial intelligences, not humans. Every time you chat with a customer service bot or ask a large language model like ChatGPT a question, you are having a conversation with a blank. There is no human thought or emotion behind their answers to your questions. Sometimes it might take a bit of going back and forth to realize that a customer service bot is not a human, but we can always tell soon enough. So the philosophical problem of other minds doesn't affect us too much in real life. We may know only our own minds from the inside, but for the most part, most of the time, we can figure out what is going on in other people's minds. This is often expressed by saying that we know about our own minds *directly*, but we have to *infer* what is going on inside other people. Our inferences about what other people think are imperfect. Linguistic and cultural differences can get in the way. Sometimes people pretend to be feeling one way when they are feeling another. And sometimes we fail, even with a close friend or family member, as when we don't realize that a comment was meant to be a joke. But most of us, most of the time, do know the minds of others. I don't speak Japanese and there are large differences between Japanese cultural norms and the North American cultural norms that I was raised with, but I can genuinely know that the other riders on the train in Tokyo are feeling uncomfortably hot: I can feel the heat and humidity; I can see them perspiring in their business wear.

Although our ideas about other minds are imperfect, we are—and should be—confident that we do know some things about them; certainly, we can always tell that the humans around us have minds. We are not in Newton Haven; we never worry whether the people around us are blanks. The problems we sometimes have in knowing what other people are experiencing never lead us to doubt that they are thinking at all. In Newton Haven, in fact, the only way to know for certain whether the thing in front of you is a person or a blank is dismemberment. We are lucky that the modern philosophical debate over the problem of other minds leaves us other options. More generally, philosophers do not worry about other humans being blanks. They also, for the most part, accept that we do have knowledge of other minds. The modern question is, How do we know about other minds?

I will argue that the problem of other minds is not really a problem and depends on a faulty understanding of what minds are. Indeed, I will also argue that the fact we think there is a genuine problem of other minds is itself a problem, one with ethical and political ramifications. To make sense of this, however, we need to look at how this faulty understanding of minds came to be and how it leads to the problem of other minds.

## THE MODERN THEORY OF MIND

René Descartes famously argued that the mind is a different sort of substance than the rest of the universe. This position is called *dualism*. The mind is known first, innately and directly;

everything else is known later and in a different way. A mind is a thing that "doubts, understands, conceives, affirms, denies, wills, refuses." The mind is not in space and does not occupy space. Therefore, it does not reflect light, weigh anything, or have any other sensible properties. Everything else—matter—does take up space and does have sensible properties. Descartes thought that the human body, like everything else made of matter, is a kind of machine. In the *Discourse on the Method*, Descartes lists the following as explicable in terms of the machinery of the human body:

1. Control of the muscles to move the body, including when the heads of the recently decapitated "still move and bite the earth."
2. The changes in the brain that produce sleep, dreaming, and wakefulness.
3. The senses.
4. Bodily feelings and other "internal passions."
5. Sensory integration.
6. Memory.
7. Imagination.

This is an impressive list. It contains a lot of the activities that we might now think of as the province of the mind, or at least the subject matter of psychology. Descartes thought he could explain all these things without invoking the mind. This would seem to leave very little for the mind to do.

This conception of the mind as unobservable thought, disembodied, and separate from sensing, moving, and the emotions leads to the problem of other minds, the worry that everyone else might be blanks. I refer to this understanding of the mind as the "modern theory of the mind," In a surprising lack of insight for such a brilliant man, Descartes does not himself notice that the modern view of the mind leads to the problem of other minds. Some of his contemporaries and correspondents, for example, the philosopher Antoine Arnauld,[1] do notice that it follows from the modern view.[2] If minds cannot be sensed but only known from the inside, how do you know that there are any besides yours?

The modern view of the mind as an inner, invisible not-body and thus the lurking problem of other minds has stuck with philosophers and psychologists ever since Descartes's time. Since then, philosophers have understood humans as deeply alone, alienated from others. For example, John Locke, whose late seventeenth-century theories were in many ways diametrically opposed to Descartes's, agreed. In his *Essay Concerning Human Understanding*, originally published in 1690, Locke argues that the invisibility of minds other than our own is the reason we have the power to speak. Although he acknowledges the problem of other minds, he is not particularly vexed by it. The vexing philosophical problem in the modern era of philosophy is the problem of skepticism, which Descartes faced head-on. Locke and other empiricists also face the problem of skepticism but acknowledge that the problem is unsolvable and that we can never really know the world around us. Locke thought that there are no innate ideas, all knowledge comes from the senses, and we can never get beyond the senses to know the independent world. Descartes thought that the information the senses provide the inner mind could be used to develop certain knowledge about the external world. Locke thought that it could not. What they have in common is that the mind is locked up on the inside, unknowable from the outside, making it so everyone but me might be blanks.

In addition to his work on the nature of the mind, Locke was an important political philosopher who influenced enlightenment thinkers like Thomas Jefferson and Jean-Jacques Rousseau.[3] His understanding of the mind informs his political views. If the vexing philosophical problem of the modern era was skepticism, the vexing political problem of the 1600s was the divine right of kings. Locke's opposition to authoritarianism connects his *Essay Concerning Human Understanding* to his *Two Treatises of Government*. First, Locke's opposition to innate ideas in the *Essay Concerning Human Understanding* is at heart an opposition to intellectual authorities:

> For, having once established this tenet,—that there are innate principles, it put their followers upon a necessity of receiving some doctrines as such; which was to take them off from the use of their own reason and judgment, and put them on believing and taking them upon trust without further examination: in which posture of blind credulity, they might be more easily governed by, and made useful to some sort of men, who had the skill and office to principle and guide them.[4]

Second, Locke defines what is generally seen as the modern version of the *self*, which is the same thinking thing that

Descartes identifies. "Self is that conscious thinking thing . . .
which is sensible or conscious of pleasure and pain, capable of
happiness or misery, and so is concerned for itself, as far as that
consciousness extends."[5] This conscious thinking self is the very
same thing as a person. "Person, as I take it, is the name for this
self."[6] This person (aka, the self; aka, the mind) is the subject
of Locke's most important political positions, positions that
cement Locke's status as one of the founders of political liberal-
ism. In the *Second Treatise of Government*, when Locke argues at
great length against the divine right of kings, he is arguing that,
in the natural state, all persons are equal and each has "perfect
freedom to order their actions, and dispose of their possessions
and persons, as they think fit, within the bounds of the law
of nature, without asking leave, or depending upon the will of
any other man."[7] This person has the right to property, which
Locke defines as "life, liberty, and estate" and must consent to
be governed.[8]

Despite being developed initially hundreds of years ago, the
modern theory of the mind is still very much with us. Although
out of fashion for the first half of the twentieth century, when
most psychologists identified as behaviorists, the modern theory of the mind was reinvigorated in the 1950s by a rightly cel-
ebrated group of computer aficionados when they founded the
disciplines of artificial intelligence and cognitive science. The
brain, they argued, is a computer, and thinking is computation;
the body is a collection of peripherals, like keyboards for input
and monitors and printers for output. These peripherals could
be safely ignored for the purposes of understanding the mind;
all that really matters is the activity of the computing brain.

This is still the view of the majority of cognitive scientists and neuroscientists, and it is hardly surprising that, as discussed in the preface, Noam Chomsky is still defending it.

The point here is that the modern theory of the mind as inner thinking thing developed by Descartes and Locke is far more politically important than most other scientific concepts. After Locke, it becomes the conception of a self or person that guides the American and French revolutions of the late eighteenth century. The modern theory of the mind is the foundation of Western liberal society. It is also a view that leaves us with the problem of other minds, one that makes others fundamentally alien to us. We can never know anyone else's minds with any kind of certainty; we can't even be certain that they have minds. The majority of this book consists of arguments—philosophical and empirical—against the modern theory of the mind.

## THE PHENOMENOLOGICAL ALTERNATIVE

Among those who reject the modern theory of the mind are twentieth-century phenomenological philosophers Martin Heidegger and Maurice Merleau-Ponty. According to these thinkers, the human mind is necessarily embodied and surrounded by technology and other people. Heidegger argued that, for the most part, people are not conscious of objects with properties.[9] To signal this, Heidegger invented a series of technical terms to describe experience. He calls humans *Dasein*, which is usually left in German in translations but would translate literally as "being there." A human, Heidegger argued, is a being-in-the-world, where the hyphenation indicates that this is a single entity. This contrasts with an observer (here) knowing facts about or having conscious experiences of material objects (over there). Instead, we are, for the most part, skillfully engaged in concrete activities in the world. The entities we engage with are not experienced as objects with properties. Heidegger signaled this with another technical term. He said that humans engage skillfully with *zeug*, a suffix in German, which translates literally as "-stuff" but is normally translated as "equipment" in Heidegger's usage of it. Equipment is plural, and it does not come in single pieces. Kitchen equipment, for example, is a connected nexus of stuff we use in the making and eating of food and the cleaning up after that making. Humans are in the world, surrounded by equipment. When we experience equipment, we experience it as usable and in terms of our abilities to use it. We are not usually conscious of the pepper grinder as a peach-colored piece of plastic, but we simply use it along with other kitchen equipment at mealtime. When we are not hungry, we see kitchen equipment (kitchen stuff) as not worthy of our attention. In both cases, we see the kitchen equipment in terms of use and in virtue of our skill at using it.

When Heidegger says that we are beings-in-the-world, the world we are in is composed of the overlapping webs of equipment we are skilled at using. We are not always skillful, however, and sometimes pieces of equipment are missing or broken. Sometimes cleverly designed pieces of equipment don't look like things that serve their function. When I was a child, for example, there was a time when phones were sometimes manufactured to look like cartoon characters or bananas or cheeseburgers. When we encounter a situation like this, we cannot skillfully use equipment. Sometimes the failure is short lived, as when we realize that it is the plastic cheeseburger that is ringing. But sometimes recovering is not so easy. If things go badly enough, we might be reduced to staring at an object with properties. When the pepper grinder is stuck or empty, I become conscious of it as a peach-colored, plastic, pentagonal prism; I turn it over and look at the metal parts on the bottom; I shake it to see if there are peppercorns inside. Outside the nexus of equipment, it is an object that I can only stare at. This staring is a degenerate mode of experience, what we do when equipment breaks down. The world that we are in is primarily experienced as usable.

This world is also necessarily social. "The world is always the one that I share with others. The world of Dasein is a with-world. Being-in is being-with others."[10] The way the networks of equipment are supposed to be used is determined by others, and every experience of equipment is also an experience of other people. We see the pepper grinder as appropriately used for seasoning food and not for killing flies or scratching an itch, although it could be used for either one because seasoning food is what other people do with pepper grinders. We could say that we see what *one* does with kitchen equipment. "By 'others' we do not mean everyone else but me—those over against whom the 'I' stands out. They are rather those from whom, for the most part, one does not distinguish oneself—those among whom one is too."[11] Recall that, according to Heidegger, most of our experience is smooth engagement with equipment, in terms of what it is used for. This is what it is to be in the world, which is to say that it is what it is to be a person at all. But what equipment is used for is determined by how others use it and understand it to be used. It follows that being a person and having experiences requires the presence of others and their experiences. Being with other people is what makes experience possible; our minds exist because theirs do as well. From Heidegger's point of view, there is no problem of other minds.

Merleau-Ponty's views differ from Heidegger's primarily because he focuses more specifically on embodiment. Merleau-Ponty distinguishes between the body as a physical object and the lived body; indeed, Merleau-Ponty's phenomenological philosophy is more than anything a theory of the lived body. Perceiving, experiencing, and thinking, according to Merleau-Ponty, are essentially embodied activities; the human mind in general is necessarily bodily, or "incarnate." "My body is the fabric into which all objects are woven, and it is, at least in relation to the perceived world, the general instrument of my 'comprehension.'"[12] The lived body opens the possibilities for action that make up the world. We do not consciously mull over these possibilities by thinking about them or imagining them; we are open to them through the skills and habits of the lived body. For example, our abilities to grasp objects make it so that we encounter objects as graspable. Our basic motor skills, as well as our more complex, culturally inflected habits, place us in an environment that consists of things to be done, objects to be manipulated. The pepper grinder appears to me the way it does in virtue of my abilities to see it, grasp it, and turn the top part while holding the bottom part over my plate.

For Merleau-Ponty, our lived bodies make experience possible, but he also argued for the malleability of the lived body. Merleau-Ponty makes this point with his long discussion of the experience of a blind person using a cane to explore the environment. The blind person's abilities and lived body change when he or she is carrying the cane so that they don't experience the cane in their hand but the world at the cane's tip. When this happens, the blind person's body schema—and therefore experience—changes. Merleau-Ponty calls this "changing our existence by appropriating fresh instruments." In doing so, he moves beyond the body to what is now called "extended mind" or "extended cognition."[13] A cognitive system is extended whenever it is partly constituted by things outside the biological body. To say that a cognitive system is partly constituted by things outside the biological body is to claim something stronger than that these things causally affect or enable cognition. Words on a page or monitor affect cognition, without being part of a thinking thing. There could be no human thought without our oxygen-rich atmosphere, but oxygen in the atmosphere is not part of us. To claim that cognitive systems are extended is to say that they are partly constituted by, that is, made up of, things outside the body.[14] In Merleau-Ponty's example, the lived body that experiences the environment is partly constituted by the cane. Merleau-Ponty takes this to imply that the boundaries of the lived body in fact are in constant flux and ambiguous at any particular moment: when the left hand explores the right, the body is temporarily part of the world; when the cane user stops exploring with the cane and puts it down, the cane switches from being part of the lived body to an object in the world.

Merleau-Ponty goes beyond the incorporation of tools into the lived body to argue that the experiences of other people are incorporated into the lived body; he calls this "intercorporeality." For example, he writes in the *Phenomenology of Perception*:

> In the experience of dialogue, there is constituted between the other person and myself a common ground; my thought and his are inter-woven into a single fabric, my words and those of my interlocutor are called forth by the state of the discussion, and they are inserted into a shared operation of which neither of us is the creator. We have here a dual being, where the other is for me no longer a mere bit of behaviour in my transcendental field, nor I in his; we are collaborators for each other in consummate reciprocity. Our perspectives merge into each other, and we co-exist through a common world.[15]

This is a theme that appears throughout Merleau-Ponty's work: mind and world and self and other are interwoven or intertwined, so much so that they together form a unified entity. This theme becomes more prominent in multiple guises in his later work. Merleau-Ponty uses a concept from Edmund Husserl, the original phenomenologist, to make these claims more forcefully.

> We need no longer explain how a being-for-itself [i.e., a self] can comprehend an other, starting from a position of absolute solitude, or how it can comprehend a pre-constituted world at the very moment that it constitutes that world. The self is inherent in the world or the world is inherent in the

self, and the self is inherent in the other and the other is inherent in the self—what Husserl calls the *Ineinander*.[16]

*Ineinander* literally means "in one another."[17] With intercorporeality and the Ineinander, Merleau-Ponty is addressing the problem of other minds directly. We do not start out, he claims, trapped on the inside, separate from others and the world. If we did, we would be in the Newton Haven situation, never certain that those around us have minds. But we do not start out trapped on the inside. We are intertwined with the world and with other people, as organs of one single intercorporeality. We are in them, and they are in us. Merleau-Ponty's view, like Heidegger's, makes the problem of other minds a nonproblem.

## Chapter Two

# THE EMBODIED MIND

**MARTIN HEIDEGGER** and Maurice Merleau-Ponty are historical figures and not cognitive scientists. In this chapter, I will look at some more contemporary approaches in philosophy and the cognitive sciences that follow Heidegger and Merleau-Ponty in rejecting the modern theory of the mind and the problem of other minds along with it.

## THE ECOLOGICAL APPROACH

James J. Gibson was a psychologist and a contemporary of Merleau-Ponty. Although they came from different traditions, both Gibson and Merleau-Ponty were influenced by gestalt psychologists, especially early in their careers. Their views are, in some respects, very similar, and late in his career Gibson found

Merleau-Ponty's work to be very valuable, so much so that in 1970 Gibson taught a graduate seminar (for *psychology* graduate students) on the *Phenomenology of Perception*. The Gibson archives at Cornell University contain detailed notes on this text. Like Merleau-Ponty, Gibson saw himself as developing a new framework for understanding perception from the ground up. Gibson's ideas are not derived from Merleau-Ponty or other phenomenologists, however, but from decades as an experimentalist, aiming to understand our perceptual experience.

We can introduce Gibson's ecological approach as consisting of three principles:

> *Principle 1:* When we perceive something, we are in direct, unmediated contact with that thing. This implies that the perceiving is not inside us but rather is part of a system that includes both the perceiving organism and the perceived object. Direct perception implies that the mind is extended in the sense discussed in chapter 1.
>
> *Principle 2: Perception is for action.* The purpose of perception is the generation and control of action, and a good deal of action is also for perception. We move our eyes, we turn our heads, we lean forward, all as part of the process of perceiving. Many proponents of the ecological approach now refuse to separate perception and action and refer instead to "perception-action."
>
> *Principle 3: Perception is of affordances.* This third principle is a consequence of the first two. If perception is direct and for the guidance of action, there must be

information sufficient for guiding action available in the environment. Affordances are properties or qualities of an object that make clear how it can or should be used, guiding action. The concept of affordances is Gibson's most widely discussed contribution.[1]

Gibson describes affordances as follows: "The *affordances* of the environment are what it *offers* the animal, what it *provides* or *furnishes*, either for good or ill."[2] A few pages later, he says,

> [A]n affordance is neither an objective property nor a subjective property; or it is both if you like. An affordance cuts across the dichotomy of subjective-objective and helps us to understand its inadequacy. It is equally a fact of the environment and a fact of behavior. It is both physical and psychical, yet neither. An affordance points both ways, to the environment and to the observer.[3]

This set of sentences is not easy to understand and has led to a cottage industry of defining and redefining the term. We don't need to go into the details of this here.[4] For now it is enough to notice that the ecological approach takes an experiencing subject to be an extended animal-environment system, including affordances for action, and not a hidden, inner entity.

Gibson was both an experimental psychologist and a phenomenologist of a sort.[5] His legacy is a robust scientific research program known as ecological psychology (a term Gibson himself never used). I will have more to say about this shortly.

## THE ENACTIVE APPROACH

The work that is viewed as the founding document in the enactive approach is *The Embodied Mind* by Francisco Varela, Evan Thompson, and Eleanor Rosch.[6] It chronicles the history of cognitive science, showing its repeated failures to capture human experience, and offers an alternative approach based on autopoietic theory, Buddhism, ideas from Merleau-Ponty, and the robots of Rodney Brooks.[7] The book also includes a specific repudiation of Gibson's ecological approach. We will see later, however, that the enactive and ecological approaches are closer than Varela, Thompson, and Rosch realized and have moved even closer in recent years.

The most clear and concise description of the enactive approach is found in Thompson's *Mind in Life*.[8] Thompson begins by pointing out that the key to understanding the relationship between experience and the material world, that is, to solving the mind-body problem, is what Merleau-Ponty called the lived body. The main question that the enactive approach attempts to answer concerns the relationship between a biological living body and a phenomenological lived body. The beginning point is that living things, starting with single cells, are self-making, autonomous systems that maintain a separation from the environment, as in the case of the cell wall. This separation is what makes a living organism an entity separate from its environment. This implies the emergence of a self, a very primitive self in the case of a cell, but a self nonetheless. The emergence of the self implies the emergence of a world, not just in that the boundary around the self leaves everything else as the world but also in that the activities of the living thing are selectively responsive to relevant things in its surround. Enactive theorists often call this sensemaking, by which they mean that the experienced world for the organism is the sense it makes of its environment.[9] Sensemaking is both cognitive and emotional, so the world is significant to the organism; it has value and valence. Sensemaking *is* cognition, at least in a minimal sense. In sensemaking, an organism maintains itself in a meaningful environment. This meaningful environment does not exist in advance of the existence of the organism but emerges along with the activities of the organism. Organisms, as lived bodies, enact or "bring forth" worlds. This biological, living body is also a phenomenological, lived body.

Hanne De Jaegher and Ezequiel Di Paolo extend the enactive approach to social interactions via what they call participatory sensemaking.[10] In participatory sensemaking, two individuals are coupled with the world and with one another so that they collectively and temporarily open a new domain of significant interactions that is not available to either separately. To take an example from De Jaegher and Di Paolo, consider what happens in what we might call the hallway dance, when each of two individual humans attempt to make space for the other while passing in a narrow hallway. Each of these individuals is an adaptive agent, engaging in sensemaking, and bringing forth a significant world. Given their interest in avoiding collision with one another, which would have negative valence for both, each will move to one side of the hallway to let the other pass. Most of the time this works perfectly well, but sometimes both individuals will move to the same side of the hallway, yielding a potential collision. Because each experiences potential collisions negatively, they each simultaneously then move to the opposite side of the narrow hallway, setting up another potential collision. To avoid the second possible collision, each individual once again switches sides of the hallway. This can happen several times, with each person mirroring the other's movement repeatedly. This hallway dance is an example of participatory sensemaking, in which each individual remains an autonomous individual, bringing forth a significant world, even while they are temporarily coupled to one another, but they also collectively bring forth a world significant to their coupled activity. Notice that, like the individuals who participate in it, the hallway dance displays its own autonomy.[11] The dance maintains itself as an entity, at least for a while, because each attempt at collision avoidance by the participants leads to another possible collision, the avoidance of which leads another possible collision, and so on.

In the case of the hallway dance, the participatory sensemaking is at odds with what is adaptive for the individuals: neither of them wants to bounce from wall to wall in the narrow corridor, but they are temporarily trapped in a social interaction. This is not always the case in participatory sensemaking, and much of social interaction also serves the individuals in the interaction. Consider (nonhallway) dancing or conversation. In these cases, a new domain of significance, unavailable to the individual participants separately, is opened by the interaction, but in this case, the participatory domain has a positive valence for the individuals. In these cases, the interaction maintains itself over time because the individual participants (dancers, conversationalists) work to maintain the interaction. In participatory sensemaking, actors are, as Edmund Husserl and Merleau-Ponty had it, "in one another."[12]

## RECONCILING THE ECOLOGICAL
## AND ENACTIVE APPROACHES

To end this chapter, I want to discuss the relationship between the ecological and enactive approaches. I have been arguing for years that ecological and enactive approaches ought to be allied.[13] This would seem natural: both reject the modern theory of the mind, and both have embraced dynamical explanation (more in part II). The approaches do differ in emphasis, however. The ecological approach focuses on the nature of the environment that animals perceive and act in; the enactive approach focuses on the organism as an agent. Combining the two would seem to provide an attractive and complete picture of cognition: an enactive story of agency, and an ecological story of the environment to which the agent is coupled. There has been recent interest in such a unification.[14] There seems to be a major philosophical barrier to such a unification: historically, the two traditions have developed from seemingly opposing starting assumptions. Ecological psychologists have traditionally asserted a commitment to realism, while enactivism was initially developed within a constructivist, and therefore antirealist, framework. Early enactivist writings can be more naturally read as advocating a form of idealism rather than realism.

The deep contrast between the two approaches is well illustrated in a remark by Varela, Thompson, and Rosch in their classic *The Embodied Mind*. The authors specifically distance their approach from ecological psychology because the latter assumes that animals live in a "pre-given" world.

> Thus the resulting research strategies are also fundamentally different: Gibsonians treat perception in largely optical (albeit ecological) terms and so attempt to build up the theory of perception almost entirely from the environment. Our approach, however, proceeds by specifying the sensorimotor patterns that enable action to be perceptually guided, and so we build up the theory of perception from the structural coupling of the animal.[15]

Around the same time, ecological psychologists critiqued the theory of autopoiesis that served as the foundation for Varela et al.'s enactivist views, calling it overtly idealist and offering a realist alternative, based in thermodynamics. Rod Swenson, for example, is extremely harsh in his criticism: "the whole concept of autopoiesis is contrived at its foundations where it is miraculously decoupled from the physical world to promulgate a solipsistic epistemology with abhorrent social consequences."[16] As is often the case, both these characterizations present straw versions of the view they critique.

These decades-old texts set up an unfortunate paucity of contact between the ecological and enactive approaches. This is primarily an educational phenomenon: ecological students read Swenson in graduate school and think that enactivists are idealists who neglect the world; enactivist students read Varela et al. in graduate school and think that ecological psychologists neglect experience. In what follows, I will argue that conceiving of the concept affordances of the ecological approach in a dynamical systems setting can help foster an easier alliance among the two approaches. Figure 2.1 depicts the interaction over time between an animal's sensorimotor abilities, that is, its embodied capacities for perception and action, and its niche, that is, the set of affordances available to it. Over developmental time, an animal's sensorimotor abilities select its niche—the animal will become selectively sensitive to information relevant to the things it can do. Over developmental time, the niche will strongly influence the development of the animal's ability to perceive and act. Over the shorter timescales of behavior, the animal's sensorimotor abilities manifest themselves in embodied action that causes changes in the layout of available

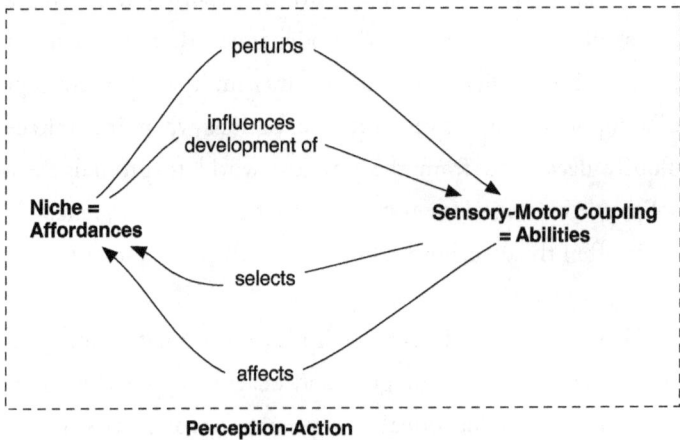

**FIGURE 2.1** The animal-environment system in *Radical Embodied Cognitive Science.* Image created by author. Anthony Chemero, *Radical Embodied Cognitive Science* (MIT Press, 2009).

affordances, and these affordances will change the way abilities are exercised in action. The key point here is that affordances and abilities causally interact in real time and are causally dependent on one another.

Figure 2.2 is an expanded version of figure 2.1 and shows the connection between organisms and sensorimotor coupling, as understood in the enactive approach. Recall that proponents of the enactive approach view the organism as a self-organizing, autonomous, autopoietic system. In this system, the nervous system generates neuronal assemblies that make sensorimotor abilities possible, and these sensorimotor abilities modulate the dynamics of the nervous system. This makes for a fully dynamical science of the entire brain-body-environment system, connecting enactivist studies of the nervous system and sensorimotor abilities to ecological studies of affordances. Of course, a diagram is not, on its own, sufficient to unify two approaches to cognition.

One reason why those who endorse the ecological approach should embrace enactivism is because of work by Rob Withagen and colleagues on the distinction between affordances and invitations.[17] At any moment, there are literally thousands of actions afforded to an agent. As I sit typing at my dining room table, the affordances include continuing to type, reaching for my glass of iced coffee, opening a browser tab to search for recent work on the Skilled Intentionality Framework, walking to the other room to pet my cat, walking to the kitchen to get a snack, walking to the kitchen to get a wet paper towel to clean up a missed spot from last night's dinner, walking to the bathroom to turn off the exhaust fan that is annoying me,

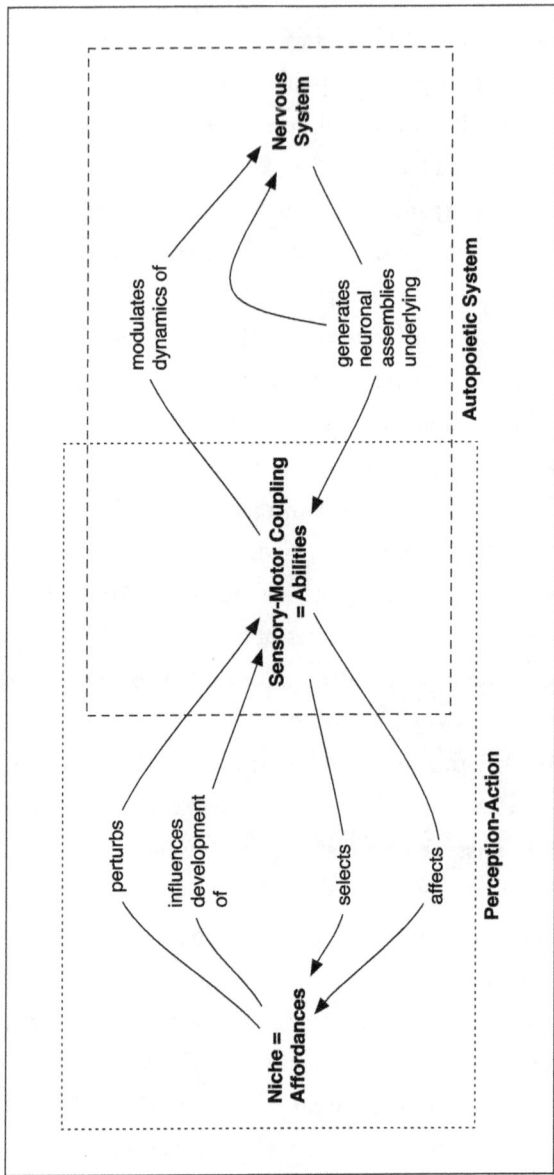

**Animal-Environment System**

**FIGURE 2.2** The animal–environment system. Anthony Chemero, *Radical Embodied Cognitive Science* (MIT Press, 2009), amended to show the animal's nervous system.

and so on, among the things I *might actually do* now. However, there are also affordances for lying on the floor under the dining table, standing on the dining table, standing on the dining chair, dancing, doing headstands, purposely spilling my glass of iced coffee, throwing my glass of iced coffee at the wall, throwing my glass of iced coffee at the window to see which would break, standing facing the corner like that creepy scene in the film *The Blair Witch Project*, standing on my head facing the corner to create a bizarre analog of that creepy scene in *The Blair Witch Project*, knocking all the cookbooks off the shelf, and so on (and on and on), among the things that I could do but that would never cross my mind unless I was trying to make a list like this. All the possible actions I just listed are affordances for me while sitting at the table typing, but only the former, shorter list of things I might actually do are also invitations. The enactive approach to agency can make sense of this distinction and of why we notice only a few of the many, many things that we are capable of doing at any moment.

For the converse, the primary reason that enactivism should embrace the ecological approach is because it is a robust, long-standing scientific research program. What I have called radical embodied cognitive science (see chapter 4) is really just a renaming of the successful scientific practices of Gibson's followers. Despite the straw position described by Varela, Thompson, and Rosch, Mog Stapleton notes that "enactivists have embraced ecological psychology," especially the idea of affordances.[18] For example, the index of Shaun Gallagher's *Enactivist Interventions* has thirty-nine listings for "affordance."[19] Gallagher uses the concept to develop theories of embodied action, imagination, and pretend play. Enactivists Ezequiel Di Paolo, Thomas Buhrmann, and Xabier Barandiaran are more guarded, but they also use the concept of affordances throughout their book and claim the enactive and ecological approaches are quite close.

> As proof of how close the approaches can be in concrete cases, we make use throughout this book of work originating in the Gibsonian tradition. This tradition usually supplies some of the clearest examples of how dynamical engagements and bodily synergies can be explanatorily powerful. But the relation between the schools of thought is one of strange familiarity, as if their respective practitioners were staring at each other across an uncanny valley.[20]

From the other side of the uncanny valley, some ecological psychologists, for example, Harry Heft, agree that the approaches cannot be fully reconciled despite their similarity.[21]

Erik Rietveld and colleagues have taken up the task of integrating the ecological and enactive approaches and expanding their reach as part of their Skilled Intentionality Framework, which understands itself as a variety of radical embodied cognitive science.[22] Figure 2.3 is an elaboration by Rietveld and Julian Kiverstein of my figure 2.2. In figure 2.3, we see the key distinction of the Skilled Intentionality Framework: that between the "field of affordances" and the "landscape of affordances available in a form of life." The former of these is the set of invitations for an individual in a particular situation; the latter is the set of all affordances for people relevantly similar to that individual. The

**RESPONSIVENESS TO A FIELD OF AFFORDANCES = SKILLED INTENTIONALITY IN CONTEXT**

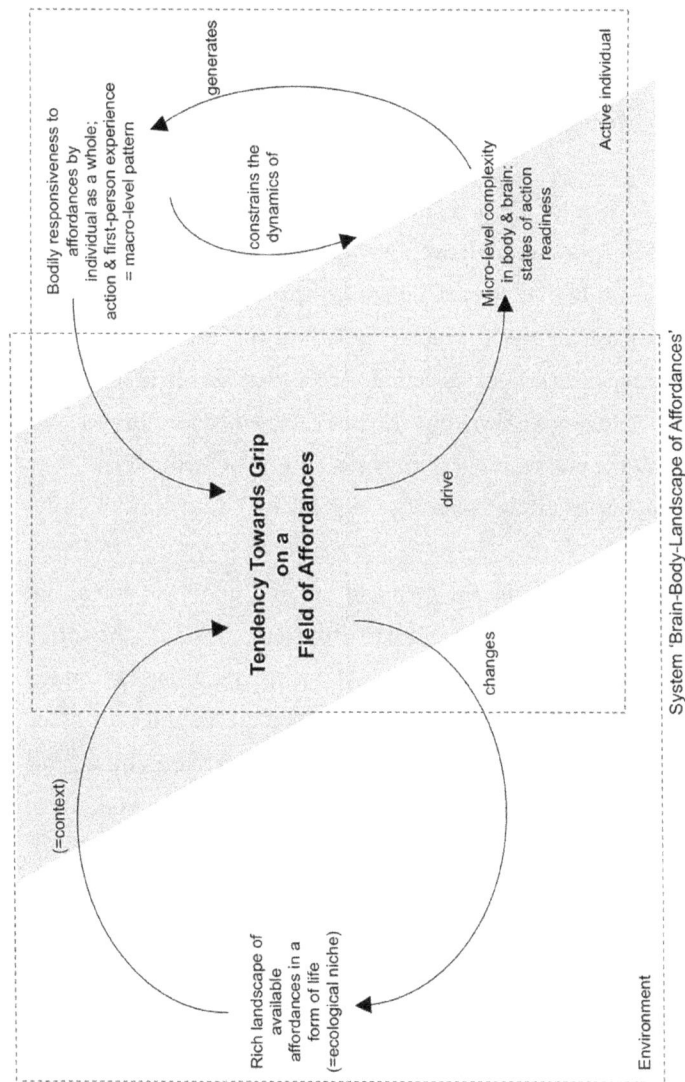

**FIGURE 2.3** The animal–environment system in Skilled Intentionality Framework. Image used by permission of Erik Rietveld and Julian Kiverstein. CCBY.

phrase "form of life," borrowed from Ludwig Wittgenstein, is important here. With it, Rietveld and Kiverstein aim to apply their ecological-enactive approach to the whole of human experience, in all its material, social, and cultural richness.

> The form of life of a kind of animal consists of patterns in its behavior, i.e., relatively stable and regular ways of doing things. In the case of humans, these regular patterns are manifest in the normative behaviors and customs of our communities. What is common to human beings is not just the biology we share but also our being embedded in sociocultural practices: our sharing steady ways of living with others, our relatively stable ways of going on.[23]

In this view, just as the forms of life of a domestic dog and a porcupine partly differ and partly overlap, so too do the forms of life of an architect and a basketball player. Given the skills they have learned in coming to participate in their forms of life, the landscape of available affordances for the architect and basketball player will both include things to eat, places to sleep, and opportunities to bathe; but only one of these forms of life includes affordances for setting a pick or dunking the ball. In a given situation, some of the opportunities will be invitations, parts of the field of relevant affordances, and others will not. At the dinner table, the affordances for setting a pick recede into the background. Within a given form of life, the key distinction here is that between the affordances available to all participants in a form of life and the invitations for an individual participant. All the basketball players on the court have acquired the skill to set picks, but only the teammates of the player with the ball are invited to exercise that skill at the moment.

The centrality of culturally learned skills in the Skilled Intentionality Framework allows us to understand the full range of human social life in terms of affordances. For social species like humans, affordances are perceived in social and cultural environments.[24] Nick Brancazio has taken this a step further, arguing that interpersonal affordances are a distinct type of affordance, characterized by the interaction of two agents. She defines interpersonal affordances as "actual possibilities for interaction with an agent," drawing on feminist and critical theory insights about how bodies can be treated as different social selves in different interactions.[25] In Brancazio's view, which affordances become invitations depends on how one is seen, that is, how one's identity is received and recognized by other agents. Political philosopher and psychiatrist Frantz Fanon describes this movingly in a chapter of *Black Skin, White Masks*, titled "Black Lived Experience" ("L'expérience vécue du noir").

I came into the world imbued with the will to find a meaning in things, my spirit filled with the desire to attain to the source of the world, and I found that I was an object in the midst of other objects.

Sealed into that crushing objecthood, I turned beseechingly to others. Their attention was a liberation, running over my body suddenly abraded into nonbeing, endowing me once more with the agility I thought I had lost, and by taking me out of the world, restoring me to it. But just as I reached the other side, I stumbled, and the movements, the attitudes, the glances of the other fixed me there, in the sense in which a chemical solution is fixed by a dye.[26]

The way Fanon is seen by others determines whether he feels himself to be an agent capable of acting in the world or a mere object.

We can see again the effects of being seen by others in this oft-cited quotation from Rosa Parks:

> "I'd see the bus pass every day . . . But to me, that was a way of life; we had no choice but to accept what was the custom. The bus was among the first ways I realized there was a Black world and a White world."

Even as a child, Parks had the bodily skills required to ride the bus to school. But the bus did not invite her to ride it. But being and being seen by others as a Black child kept opportunities to ride the bus to school out of her field of affordances, even though the bodily skills to take advantage of the opportunity to ride the bus are available in the human form of life. For humans in our sociocultural forms of life, how we are seen by others determines how we see ourselves and what we feel capable of doing.[27]

Seeing affordances in this ecological-enactive light makes clear that ideas from the cognitive sciences can have deep political import. This will be important later this book. In the next chapter, we will see how the ecological-enactive approach supports an intertwined view of the self.

*Chapter Three*

## THE INTERTWINED SELF

ACCORDING TO the modern view of the mind, which John Locke and René Descartes share with most cognitive scientists, the mind is a conscious thinking thing, hidden and inner, separated from the rest of the world. Recall that Locke defines "self" and "person" in the very same terms. We have seen some philosophical pressure put on this understanding of the mind by phenomenological philosophers and proponents of the ecological and enactive approaches. In the chapters that follow, we will put some scientific pressure on this view of the mind (and the self and the person). In this chapter, we will see how proponents of the ecological and enactive approaches understand the self. Why is the focus changing from mind to self? Despite Locke's definitions, recent philosophical practice has been to view self, mind, and person as not identical. A person is thought generally to contain a mind (usually, plus a body and some history); a mind is thought to contain the self as a proper part at its core.

The self is generally understood as the nugget in you that makes you the same person over time. When we see the self reconceived as necessarily embodied and social, we will also see the mind and person as necessarily embodied and social.

## ECOLOGICAL APPROACHES TO THE SELF

James Gibson studied perception, so he focused on the perception of the self. Perception of the self was originally understood as depending on a separate sense called proprioception. The classical understanding of proprioception comes in the early twentieth century from Charles Sherrington, who also coined the term "synapse."[1] Sherrington distinguished between proprioception and exteroception. Exteroception is the job of the five senses—seeing, hearing, touching, smelling, tasting—which work in the service of exteroception, or perception of the world surrounding the perceiver. Proprioception, which is also called kinesthesis, is the perception of the locations and movements of parts of the body. Sherrington thought that the basis of proprioception was a specialized set of cells called proprioceptors that are arrayed at muscles and joints and are sensitive to position and movement. In contrast, Gibson argued that proprioception was not a special sense and that all the senses were both proprioceptive and exteroceptive.

To make sense of this, we can start by defining the term "field of view." Sit very still and look straight ahead. When you do so, a portion of the light in the room is reflected to your eyeballs. This is your field of view. Your field of view is determined by the shape of your head and the placement of your eyes on it. Humans have forward-facing eyes, so the backs of our heads block a large proportion of the light in the room. This is seen in the famous self-portrait by German mathematician, physicist, physiologist, and philosopher Ernst Mach, which he called "view from the left eye" (see figure 3.1).[2] As you can see in the figure, about a third of the room is visible vertically, with some of the light is blocked by our face and eyebrows. With two eyes and from a horizontal perspective, the field of view excludes

**FIGURE 3.1** Self-portrait, or view from the left eye, by Ernst Mach. Public domain.

slightly less than half of the room. The key point here is that the information in your visual field of view enables perception of the environment, the book or the screen you are holding in front of you, the painting on the wall behind it, the cat entering at the periphery. But your field of view also allows you to perceive yourself. You can see the shape of your nose, the placement of your eyes on your face, the shape of your face. What you could see would be different if your nose was larger or your eyes were set slightly more widely or if you were taller. The specifics of your body make what you see different from what other people see. Gibson makes a starker comparison.

> The horse and human look out on the world in different ways. They have radically different fields of view; their noses are different, and their legs are different, entering and leaving the field of view in different ways. Each species sees a different self from every other. Each individual sees a different self. Each person gets information about his or her body that differs from that obtained by any other person.[3]

Seeing is always seeing both yourself and the world.

The inseparability of proprioception and exteroception is even more plain when an agent moves. This is hinted at in the quotation above. As I walk, I swing my arms and they go in and out of my field of view. This is a small example of what Gibson called visual kinesthesis, the visual perception of one's own movement. Pick an object that is about at the height of your eyes, say, a painting on the wall, and walk toward it. Notice that, as you get closer to the painting, the proportion of your field of view that it occupies increases. When you get close enough, with the painting just a few centimeters from your nose, it takes up your whole visual field. This increase, called optical expansion, is nonlinear. At ten meters, moving one step closer to the painting causes a smaller increase in the visual field than moving a step closer from two meters away. It is also reversible and leads to nonlinear optical contraction, in which moving away from something that is very close causes the proportion of the visual field to shrink more quickly when the painting is close than when it is further away. This optical expansion and contraction are types of visual information about your motion and enable visual kinesthesis. Visual kinesthesis has been studied in great depth by ecological psychologists. The most striking version of this work, by David Lee and Eric Aronson, shows visual kinesthesis in toddlers using a moving room: a room with movable walls suspended just above the floor.[4] Lee and Aronson showed that suddenly moving the walls made the toddlers fall in a predictable way. If they moved the wall suddenly toward the toddlers, they had the same kind of optical expansion that they would have had if they were falling forward, causing them to correct toward the back and to fall backward; suddenly moving the wall away from the toddlers gave them the kind of optical contraction that would occur if they were falling backward, causing them to overcorrect toward the front and fall forward. The artificial optical expansion and contraction caused the toddlers to see themselves falling. In a later study, David Lee and Roly Lishman showed that the same happens to precariously standing adults.[5] That is, if you were asked to stand on a balance beam, the moving wall would make you fall off in the predicted direction. This is visual kinesthesis, the perception of the self and actions without any special, separate proprioceptive sense or sensors. The point here is that perceiving the world is also always perceiving the self.

Gibson did not pursue a theory of the self beyond the perception of the self, but other ecological psychologists did. Ulric Neisser, for example, worked out ecologically inspired approaches to memory and the self. Neisser developed theories of five different kinds of self-knowledge and five corresponding aspects of self.[6] It is important to keep in mind that, although we will follow Neisser in calling each aspect a kind of self, the claim is not that any individual is more than one kind of self. In the typical case, we have one self, which has many aspects. What Neisser called the ecological self is the self as perceived in visual (and auditory and so on) kinesthesis, and Neisser acknowledges that it is essentially James Gibson's idea. Ecological selves are directly perceived simultaneously with the direct perception of the world that surrounds them, including "among other things: where they are, how they are moving, what they are doing, and what they might do, whether a given action is their own or not."[7]

Neisser's interpersonal self is also present from birth. This self is directly perceived during interactions and communications with others, including during in utero interactions with others. When two people engage in any dyadic interaction, they are each aware of themselves as interpersonal selves, as participants in an ongoing social interaction that they are working to maintain in concert with the other. Each experiences their interpersonal self as cocreator of the ongoing interaction and as the focus of the other's attention. Just as with the case of eco-logical self and the world, the perception of the interpersonal self is always simultaneous with the perception of another.

Human beings confirm one another's selfhood in so many ways that it is impossible to list them all. Almost every personal encounter is mutually regulated: A directs behavior toward B, and B to A, in a reciprocal pattern that both establish together and both perceive. This pattern exists objectively and observably.[8]

The interpersonal self is also present from infancy and can be seen in the close attunement of emotion and mood between infants and caregivers. Nursing is a social engagement in which the mother and child both see themselves and one another as participating in a social interaction, an interaction that they cocreate and comaintain.[9] Most interactions between infants and caregivers are interactions between interpersonal selves.

Neither the ecological nor the interpersonal self is hidden and inner. At the same time, they are not intended to account for some kind of nugget that makes you who you are. From the perspective of the ecological approach, this is the case because there could not be such a nugget. You experience your ecological and interpersonal selves, which are objective perceivable things, whenever you move around the world or interact with another human. Neisser also discusses three other kinds of self, which are developmentally later and not directly perceivable. The conceptual self is a set of things with which you identify yourself. I am a father, a husband, a cognitive scientist, a Soul

Glo fan, a (poor) karateka, and so on. The extended self is your memory for events that happened in the past. The private self is your stream of conscious experience, including experiences of things that are distant in space and time. These three selves are not perceivable features of the environment for you or anyone else; they are private and depend on memories that you have stored. Neisser points out that, for most of the history of Western philosophy, the private self has been the only self that people have discussed. The private self is the mind according to modern theory of the mind. Neisser's ecological approach to the self does not deny that the private self exists. However, the private self does not have the developmental or epistemological priority that the modern theory ascribes to it.

## PHENOMENOLOGICAL AND ENACTIVE APPROACHES TO THE SELF

Unlike proponents of the ecological approach, phenomenological and enactive theorists have been very interested in the nature of the self, so much so that I cannot hope to do justice to the range of views here. We can start, however, with what has become the common distinction between the *minimal self* and the *narrative self*. Shaun Gallagher, who is both a phenomenologist and an enactivist, defines the minimal self as follows: "Phenomenologically, that is, in terms of how one experiences it, a consciousness of oneself as an immediate subject of experience, unextended in time."[10] This minimal consciousness of oneself seems to include a *sense of agency* and a *sense of ownership*.

The sense of agency is the experience of being a purposeful actor, the feeling of being a *doer*; the sense of ownership is the feeling that experiences are *mine*. The senses of ownership and agency are bound tightly together most of the time. I feel myself violently stomping the tail of the skateboard onto the pavement with my back foot and immediately jumping while scraping my front foot along the length of the board. I am the one who did the Ollie, and I am the one who felt myself become briefly airborne before pushing the board back down onto the pavement. But the senses of agency and ownership can also be felt separately. As is often the case when I have tried to do an Ollie, I feel the board scooting out from under my feet and feel myself falling toward the pavement; the falling is happening to me but it is not something I am doing. The distinction between the senses of agency and ownership has been conceptually and empirically fruitful.[11] It is also controversial. Sanneke de Haan and Leon de Bruin argue that the sense of agency and sense of ownership are deeply intertwined, so much so that they cannot be viewed as separate senses.[12] Dan Zahavi has made the minimal self more minimal still, arguing that the most minimal notion of self doesn't require a sense of agency.[13]

In contrast, the narrative self is "[a] more or less coherent self (or self-image) that is constituted with a past and a future in the various stories that we and others tell about ourselves." The narrative self derives from Daniel Dennett's earlier work on the self as the "center of narrative gravity."[14] Dennett's antiphenomenological idea is that there is no such thing as a minimal self; there is nothing to a self beyond our tendencies to tell stories aloud and sotto voce about a particular character with a particular history, without the need for an unmeasurable sense of agency or ownership. For Dennett, the former is just our tendency to describe ourselves as doing things on purpose; the latter is just the tendency to describe things as happening to us. With his pattern theory of the self, Gallagher argues that we need multiple senses of self, including the minimal self and the narrative self, to do justice to the phenomenon. For Gallagher, as for Daniel Hutto, many aspects of our experiences and cognitive abilities are wrapped up in human narrative practices.[15] Learning to understand stories that others tell about their thoughts, feelings, and desires and learning to tell similar stories about ourselves are partly what enables us to have those thoughts, feelings, and experiences.

It might seem natural to align the minimal self with Neisser's ecological self. Both the minimal and ecological selves point to the ways we own our experiences as they happen. I see things from a particular pair of eyes on a particular head on a particular body in a particular environment; *I* am the one seeing these things. There is a key disconnect, however, stemming from the claim that the minimal self is "unextended in time." This is inconsistent with both ecological self and with the discussion in chapter 2 of the enactive approach. The ecological self is experienced in kinesthesis. Moving is not unextended in time. For the enactivist self, the distinction between the self and its environment is in constant dynamical flux; that is, it is a process. Processes are extended in time, by definition; they are not metaphysical nuggets. In fact, no experience, indeed nothing in nature, is instantaneous. Going forward, then, we will strike the phrase "unextended in time" from the definition of the minimal self. The senses of agency and ownership are extended in time, as is the minimal self. Zahavi, who has become the most vocal proponent of the minimal self, agrees.[16] Along the same lines, we can think of the narrative self in terms of Neisser's conceptual, temporally extended, and private selves. The center of our narrative gravity has a series of properties and is often described using concepts applied to ourselves. It depends on memories. As noted above, we can tell ourselves stories in which we star quietly without speaking anything aloud.

What about Neisser's interpersonal self? On its face, it is neither an aspect of the minimal self nor of the narrative self. Neisser argued that, like the ecological self, the interpersonal self is directly perceivable and present from birth but that it depends on different (unspecified) mechanisms. How then does it fit with phenomenological and enactive understandings of the self? Gallagher thinks the interpersonal self sits alongside the minimal and narrative aspects of the self. It is not alone: in his article, Gallagher discusses eight different aspects of the self.[17] Zahavi also thinks the interpersonal self is separate from the minimal and narrative selves.[18] In contrast, Sanneke de Haan, Miriam Kyselo, and Anna Ciaunica argue (separately) that there is no minimal self that is not also interpersonal.[19] De Haan draws on phenomenology and developmental psychology.[20] With Martin Heidegger, de Haan claims that every experience that we have depends on the presence of others and their experiences. In Heidegger's way of putting it, the world we experience is a with-world. Maurice Merleau-Ponty makes the same point eloquently.

> To begin with [other persons] are not there as minds, or
> even as "psychisms," but such for example as we face them
> in anger or love—faces, gestures, spoken words to which our
> own respond without thoughts intervening, to the point that
> we sometimes turn their words back upon them even before
> they have reached us, as surely as, more surely than, if we had
> understood—each one of us pregnant with the others and
> confirmed by them in his body.[21]

Infants are social from birth and arguably before then. De Haan
invokes Vasudevi Reddy's arguments[22] that infants' first experiences of themselves are as objects of the attentions of others.
This can be seen very early in typically developing infants, who
seek the attention of others from birth. From the beginning,
we are pregnant with others. De Haan concedes that it might
be possible to distinguish conceptually between the minimal
self and the interpersonal self, as Neisser and Zahavi and Gallagher do, but argues that they are equiprimordial and equally
the products of development. It is just not the case that the
interpersonal self is a later addition, dependent on the prior
existence of the minimal or ecological self. The minimal and
interpersonal selves are like Batman and Bruce Wayne—a single thing that goes by more than one name.

Anna Ciaunica also makes a developmental and phenomenological argument for the sociality of even the minimal
self.[23] Ciaunica intends to counter Jean-Paul Sartre's view of
the relationship between being-for-itself and being-for-others.
Being-for-itself is the mode of being of an experiencer. Every
being-for-itself has prereflective self-consciousness, what we

have been calling a minimal self. In contrast, being-for-others is a kind of alienation: my being-for-others is experiencing myself as a separate person who can be seen by a distal other. Being-for-itself is a given for anything that has experiences; being-for-others, for Sartre, is a later achievement. Ciaunica focuses on both early life and in utero experiences to argue that this not how these two modes relate. Although not everyone becomes pregnant, everyone began life inside another human. From the very beginning, we are coembodied, coenactive, cometabolizing with our mother. All in utero experiences are experiences with another person; they are participatory sensemaking.[24] This does not suddenly change when an infant is born; early experiences typically involve bodily coupling with a caregiver. The infant is touched by and is touching a caregiver. The experience is not of being an object that is experienced by another; rather, it is the experience of a cosubject in coawareness. For the infant, prereflective self-awareness, that is, the minimal self, is experienced first and foremost as other-related. For Ciaunica, the minimal self is relational in nature.

Miriam Kyselo reaches a similar conclusion from an enactivist starting point.[25] She starts with the tension inherent in the enactivist approach between the adaptive living system, which is constantly maintaining itself as separate from the environment, and its need to be structurally coupled to the environment. Staying alive is always tenuous and requires a balance between connection to and separation from the environment. There is a tension, that is, between self-maintenance and sensemaking. As we saw above, the environment that humans make sense of includes other humans from the very beginning.

The upshot of this is that sensemaking is always at least partially participatory sensemaking. Kyselo sees these two activities in a perpetual tug of war, with our needs to individuate ourselves from the world and to maintain ourselves as living bodies constantly at odds with our need to engage in sensemaking, especially participatory sensemaking. She calls this the body-social problem—humans need simultaneously to maintain themselves as living bodies, potentially at the expense of participation, and engage in participatory sensemaking, potentially at the expense of self-maintenance. The solution to the problem is to acknowledge the tension, even as an aspect of the minimal self. As she puts it in the title of an article, "the minimal self needs a social update."[26]

We have looked at the self from several different perspectives: the ecological approach, developmental psychology, the enactive approach, phenomenology. From each of these points of view, we have seen that the minimal self is always already social.

## SENSORIMOTOR EMPATHY AND DIRECT SOCIAL PERCEPTION

In the modern theory of the mind, knowledge of the minds of others is possible only by inference, and quite sophisticated inference. We have to see how others are behaving and use what we see to theorize about or simulate their experiences. Because we are generally not very good at theorizing, the problem of other minds arises: what if my theorizing about your mind is hopelessly mistaken and you are just a blank, a sophisticated automaton? But we have just seen that even infants, from birth, are aware that their caregivers experience things along with them: they are touched and touching; they are copresent in coexperiences. I want to suggest that the way to characterize copresence in coexperiences is in terms of a kind of empathy. As Joel Krueger points out,[27] "empathy" does not have a long history as an English word: American psychologist E. B. Titchener introduced the word as a translation of psychologist Wilhelm Wundt's German *Einfühlung*, literally, "feeling into." The German word had been used originally to describe the feelings invoked by works of art. Only later was the usage of the term expanded to refer to feeling-into other humans, the phenomenon now more commonly called empathy. The word "empathy" has indeed come a long way from the initial sense of feeling-into. In current usage by psychologists and philosophers of mind, "empathy" refers to (1) knowing that another person is having feelings, (2) knowing what those feelings are like, and (3) having appropriate feelings in response to those feelings. In this latter sense, empathy has become a central concern of social psychology. Empathy in this sense, however, is not relevant to understanding the experiences of infants. First and foremost, empathy is intellectual, not sensory; it is having feelings in response to knowing about the feelings of others. The original sense of empathy, of einfühlung or feeling-into, however, is relevant to infant experience. In effect, before they are able to simulate or theorize, infants feel-into their caregivers, experiencing them as responsive to their explorations. I have called this feeling-into "sensorimotor empathy,"[28] in contrast with the cognitive empathy that social psychologists and philosophers typically discuss.[29] Even as adults, we engage in sensorimotor empathy when our lived body expands and temporarily includes aspects of the nonbodily environment, whether those aspects are tools or other humans. Sensorimotor empathy is skillful, implicit, and bodily engagement; it is crucial for making sense of the experiences of both adults and infants.

Sensorimotor empathy is experiencing yourself as expanding to include other things. Sensorimotor empathy differs from the cognitive form of empathy described above in several ways. First, it is more basic to our engagement with the world than cognition or emotion would be. It is what Giovanna Colombetti calls affective, to refer to basic, unreflective, valenced experiences.[30] Affect is the foundation on which (more) sophisticated cognitive or emotional activities stand. Sensorimotor empathy is genuinely sensory and not dependent on explicit thoughts or concepts or abilities to simulate or theorize. Someone who navigates the world with a cane doesn't think about the cane; they feel the world through it. Dancers don't think about what their partners are doing; they engage in constant push and pull so that they form a unit. Engaging in sensorimotor empathy is not a matter of knowing about a tool or another human. It is the engagement of the ability to connect. Sensorimotor empathy is a matter of genuine, two-way engagement. While watching a movie, you can feel cognitive empathy for a character because you know that they have been dumped by the person of their dreams, how you would feel if you were dumped by the person of your dreams, and you have an appropriate emotional response. However, the movie character does not respond to your empathy. In contrast, a dance partner responds. This is the very definition of being a dance partner. You can feel into a dance partner but not a movie character; a caregiver and infant feel into one another. This sensorimotor empathy constitutes our most basic ways of being with one another and being in the world. To echo Kyselo, the minimal self needs a social update precisely because infants engage in sensorimotor empathy from the very beginning. There is no self that is not with others.

The understanding of the necessarily social self that I have been developing here aligns with Gallagher's interaction theory of our knowledge of other minds.[31] According to interaction theory, we do not need to make inferences to know the minds of others in most circumstances; we simply interact with them. He sometimes puts this in terms of "direct social perception."[32] Just as ecological psychologists argue that we can perceive the environment directly, without intervening inferences, Gallagher argues that we perceive the experiences of others without intervening inferences. I would phrase this by saying that we engage in sensorimotor empathy. We are intertwined with those we engage. Sensorimotor empathy sometimes fails us. When there is deception or misunderstanding or language barriers, we can no longer feel into one another. When that occurs, we can engage in simulation of or theorizing about what the other is experiencing. The ability to do so appears at around three years of age, along with other conscious reasoning abilities. In contrast, the sensorimotor empathy that enables the interpersonal self is always present.

## SUMMARY: THE INTERTWINED SELF

In these first three chapters of this book, I have described two contrasting views of the mind, self, and person. According to what I have been calling the modern theory of mind, the self is a proper part of the individual mind that is a proper part of an individual person. According to the modern view, these are all hidden and inner, so much so that it is easy to imagine that we are systematically wrong about the goings on in the minds of others. This view has structured our philosophical, scientific, and political approaches since the Enlightenment. It also leaves us in Newton Haven alongside the characters from *The World's End*, with the problem of other minds. I have also developed a view based on ideas from phenomenology and the ecological and enactive approaches, according to which even the core minimal self is interwoven with the world and with others. According to this view, which I will call the intertwined self from now on, minds are mostly not hidden or private. According to this view, being a person is first and foremost being with others from whom our minds are not hidden and some aspects of whose minds we can directly perceive. There are no blanks in the intertwined view. Even our core selves are enmeshed in the core selves of others.

I strongly prefer the intertwined view. But nothing I have said so far has amounted to an argument for it or against the modern view. The job of the next two chapters will be to develop an empirical case for the plausibility of the intertwined self.

# PART II

*Every human being is a puppet on strings, but the puppet half controls the strings, and the strings do not ascend to some anonymous Maker, but are glistening golden strands that connect one puppet to another. Each strand is sensitive to the vibrations of every other strand. Every vibration sings not only in the puppet's heart, but in the hearts of many other puppets, so that if you listen carefully, you can hear the strum of many hearts singing together . . .*

Jeff Vandermeer, *Ambergris 2*

*Chapter Four*

# RADICAL EMBODIED COGNITIVE SCIENCE

## A NOTE ON CHAPTERS 4 AND 5

This chapter and the next involve some technical material, which I have suppressed in the main text. The discussion in these chapters is therefore, in a sense, incomplete. For those who want the mathematical details, I have included an appendix at the end of the book and pointers to specific sections in the appendix here in the main text.

## INTRODUCING RADICAL EMBODIED COGNITIVE SCIENCE

The purpose of this chapter is to characterize an attempt to describe a scientific psychology that rejects the modern theory of the mind and can be used to gather evidence for the intertwined self, as described in chapter 3. Current mainstream cognitive science is poorly suited for this purpose. The cognitive revolution was, in effect, a reinstallation of the modern theory of the mind, after several decades of dominance by behaviorism. As usual, the philosopher Jerry Fodor describes it well:

> The trick is to combine the postulation of mental representations with the "computer metaphor." . . . In this respect, I think there really has been something like an intellectual breakthrough. Technical details to one side, this is—in my view—the only aspect of contemporary cognitive science that represents a major advance over the versions of mentalism that were its eighteenth- and nineteenth-century predecessors.[1]

Current mainstream cognitive science, that is, is the modern theory of the mind, plus the idea that the brain is a computer. Since the cognitive sciences really took hold in the 1970s and 1980s, the psychological mainstream has adopted the modern theory of the mind. For the most part, Western culture has also adopted the view that the mind is a set of computational processes that happen deep in the recesses of the thinker's brain, hidden from outsiders. You can see this in the (admit it: terrible) *Matrix* films and in *The World's End*, the much better film that frames this discussion.

The brain in a vat is a long-standing thought experiment according to which we are disembodied brains, floating in some kind of fluid and controlled from the outside by a super-computer or evil neuroscientist so that our experiences are as if we are embodied and engaging with a real world. The claim is that we cannot know that we are not brains in vats. The possibility that we can be brains in vats is aligned with the modern theory of the mind and with cognitive science, and can be seen in another claim made by Fodor, namely, that scientific psychology needed to adopt *methodological solipsism*.[2] Because our experiences are nothing more than computational processes in our brains, we should study experiences by focusing only on those computational processes, without concern for the world the processes seem to be about. This is worse than Newton Haven: not just might your best friend be a blank, but the concert that you think you and your friend are seeing together might not really be happening. The idea that we might be brains in vats might seem intuitively plausible, but it has been decisively refuted many, many times. Yet the concept of the mind as hidden and inner that underlies it is still central in the cognitive sciences. Phenomenological approaches, the ecological and enactive approaches, and the intertwined understanding of the self that follows from them require a different scientific approach. I have called this approach radical embodied cognitive science. The name comes from Andy Clark, who coined the term "radical embodied cognitive science" to label a view he took to be unreasonable.[3] The psychology, to be described in this chapter, combines nonlinear dynamical modeling with ideas about the nature of the mind from the ecological and enactive approaches.

To see how radical embodied cognitive science works, consider a theory of the relationship between mind and world as general as those by James Gibson, phenomenologists such as Martin Heidegger or Maurice Merleau-Ponty, and proponents of the enactive approach. In each case, the theory begins with a critique of an understanding of cognition according to which the world causally impinges on the thinker and causes the thinker to form internal representations of the world. These representations are the thinker's only access to the world. This, of course, is the modern theory of mind, hidden and inner. With the exception of a few decades in the twentieth century, this has been the scientific and lay understanding of the mind, from René Descartes to today's cognitive science. This picture was explicitly criticized by Heidegger, Merleau-Ponty, and Gibson. As Heidegger puts it, "[T]he perceiving of what is known is not a process of returning with one's booty to the 'cabinet' of consciousness after one has gone out and grasped it."[4] In contrast, in the world Gibson and the phenomenologists depict, a thinker is surrounded by and interacting with the world itself and need not form representations of it to do so.[5] Very general theories of the relationship between thinking and the world are instructive and inspiring but can be difficult to put into contact with data gathered in the lab. These very general pictures of the nature of the mind are not in themselves scientific research programs. To make them into scientific research programs and to put them into contact with data you can gather in the lab, something more is needed. In the case of radical embodied cognitive science, that something else is dynamical systems modeling.

## THE DYNAMICAL SYSTEMS
## APPROACH TO COGNITION

The recent interest in dynamical modeling in the cognitive sciences begins with work by Peter Kugler, Scott Kelso, and Michael Turvey, specifically with their attempt to answer a question raised by Gibson. In trying to explain action in a way that did not demand an inner agent using sensory representations of the world to develop motor representations of the actions to be taken, Gibson says, "The rules that govern behavior are not like laws enforced by an authority or decisions made by a commander: behavior is regular without being regulated. The question is how this can be."[6] Kugler, Kelso, and Turvey answered this question by arguing that human behavior is self-organizing and therefore ought to be modeled by the mathematical tools that one applies to self-organizing systems in other sciences. A self-organizing system is one that exhibits regularities that arise without a plan or leader but from the interactions of the parts of the system. Self-organizing systems can exhibit these regularities by using energy from their surroundings to create patterns. To take a mundane example, consider the whirlpool pattern that forms when a toilet is flushed. The water molecules temporarily move in a coordinated way that is very atypical for water molecules. They can exhibit this pattern of behavior because potential energy from the water in the tank is released when the toilet is flushed. This energy enables the pattern to form and last until the energy is dissipated, after which the water molecules go back to their more typical behavior.

In physics, systems that use energy are explained using dynamical systems theory. Kugler, Kelso, and Turvey brought these methods into the cognitive sciences.

A dynamical system is a set of variables changing continually, concurrently, and interdependently over time in accordance with dynamical laws. In principle, if not always in practice, these laws can be described by a set of differential equations that model how the variables change over time. Dynamical systems theory is especially appropriate for phenomenological, ecological, and dynamical approaches because single dynamical systems can have parameters and variables both within the agent and in the environment or another agent. In a classic article, Randy Beer describes a clear conceptual approach to dynamical cognitive science, including offering placeholder equations (see figure 4.1).[7] Beer suggests that we might explain the behavior

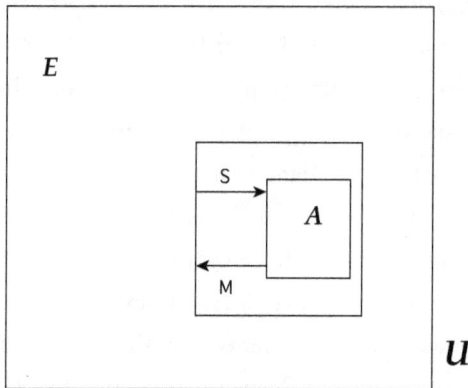

**FIGURE 4.1** The agent-environment system. Redrawn by the author from Randy Beer, "A Dynamical Systems Perspective on Agent-Environment Interaction." *Artificial Intelligence* 72, no. 1–2 (1995): 173–215.

of the agent in its environment over time as coupled dynamical systems, using something like the following equations:

$$\dot{X}_A = A\left[X_{A;}S\left(X_E\right)\right]$$
$$\dot{X}_E = E\left[X_{E;}M\left(X_A\right)\right]$$

where $A$ and $E$ are dynamical systems, modeling the organism and its environment, respectively, and $S(X_E)$ and $M(X_A)$ are coupling functions from environmental variables to organismic parameters and from organismic variables to environmental parameters, respectively. The dots over the variables on the left side indicate that these are differential equations, which model how things change over time. The first of these equations says that the way an agent changes its behavior over time ($\dot{X}_A$) is a function ($A$) of how the agent is now ($X_A$) and how it is sensing the environment now [$S(X_E)$]; the second equation says that the way the environment is changing over time ($\dot{X}_E$) is a function ($E$) of how the environment is now ($X_E$) and what the agent is doing now [$M(X_A)$]. In plainer English, these equations say that the way an agent changes depends on how it is now and how it is sensing the environment, and that the way the environment changes depends on how it is now and on what the agent is doing. For example, seeing the world is a coordinated dance between eye movements and light: eye movements are partly caused by the light that hits the eyes; what light hits the eyes is partly caused by eye movements.

The key point here is that these equations can be solved only at the same time. We would say in mathematics that the equations are *coupled*. The changes in the agent's action depend on how the environment is and changes in the environment depend on the how the agent is acting. To Beer, this suggests that the agent and the environment form a single coupled system, $U$. What needs to be explained is the way $U$, the coupled agent-environment system, unfolds over time. The fact that the mathematical model works this way reveals something important about the relationship between animals and their environments: you cannot make sense of them separately.

The use of dynamical systems theory as a modeling tool plays several crucial roles in radical embodied cognitive science. First, and perhaps most important, is that it does what modeling does throughout the sciences: it bridges the gaps between abstract theorizing and concrete data that can be gathered in the lab. Second, the alternative to the modern theory of the mind that I have been developing requires an understanding that does not take the mind to be hidden and inner. Phenomenological, ecological, and enactivist approaches demand an explanatory tool that can span the agent-environment border. Dynamical systems explanations do this. We will see that dynamical systems explanations can also span that agent-agent border.

What I am calling radical embodied cognitive science then is the use of dynamical modeling to put the theoretical positions of phenomenology and the ecological and enactive approaches in touch with data about perception, action, and cognition, which can be gathered in the lab. People have been doing radical embodied cognitive science, under different names, for decades now.[8] In the next section, we will see how this works in practice.

## COORDINATION DYNAMICS

The locus classicus for dynamical research in cognitive science is work from the 1930s by Erich von Holst.[9] Von Holst studied the coordinated movements of the fins of swimming fish, which he modeled as coupled oscillators. Coupled oscillators had been studied by physicists since the seventeenth century, when Christiaan Huygens described the "odd sympathy" of pendulum clocks against a common wall. Clocks like this are physically connected to one another via the wall, making the back and forth of their pendula connected such that they consistently approach but never quite fall into synchrony. Von Holst assumed that a fish's fin, on its own, has a preferred frequency of oscillation. When coupled to another fin via the fish's body and nervous systems, each fin tends to pull the other toward is preferred period of oscillation and also tends to behave in a way that maintains its own preferred period of oscillation. These are called the magnet effect and the maintenance effect, respectively. Von Holst observed that, because of these interacting effects, pairs of fins on a single fish behaved like clocks against a common wall: they tended to reach a cooperative frequency of oscillation, but with a high degree of variability around that frequency. This activity can be modeled with a simple pair of coupled equations, in which the activity of each fin depends on its own preferred period of oscillation and the effects on its behavior caused by being coupled to the other fin via a body and nervous system (See the appendix, section 1.)

This basic model is the foundation of later coordination dynamics, especially in its elaboration by Hermann Haken,

Scott Kelso, and Herbert Bunz.[10] The Haken-Kelso-Bunz (HKB) model elaborates the von Holst model by including concepts from the physical properties of coupled oscillators as a way to explain self-organizing patterns in human bimanual coordination. That is, it uses resources from synergetics, a twentieth-century innovation in physics designed to deal with far-from-equilibrium physical systems like flushing toilets and living things. Synergetics will be discussed more in chapter 5.

In earlier experiments, Kelso showed that human participants, asked to tap the index fingers of both of their hands along with a metronome, invariably tapped them in one of two patterns.[11] Either they tapped the fingers in phase, with the same muscle groups in each hand performing the same movement at the same time, or out of phase, with the same muscle groups in each hand "taking turns". Kelso found that out-of-phase movements were stable only at slower metronome frequencies; at higher rates, participants could not maintain out-of-phase coordination in their finger movements and transferred into in-phase movements. (For details of the model, see appendix 1, section 1.)

The HKB model accounts for the behavior of this two-part coordinated system. Given the model's basis in synergetics, it makes a series of other predictions about finger wagging, especially concerning system behavior around critical points in the system's behavior. All these predictions have been verified in further experiments. The HKB model does not apply just to finger tapping. In the years since the original HKB article was published, it has been extended many times and applied to a wide variety of phenomena in the cognitive sciences, from limb coordination to brain-body coordination, to the coordination between brain areas. We will shortly discuss that it also applies to interpersonal coordination.[12] While coordination dynamics has been the most visible of the fields of application of dynamical models in the cognitive sciences, it has hardly been the only one. Dynamical modeling has been applied to almost every phenomenon studied in the cognitive sciences. It is well established and it is here to stay. It will also be the key to finding empirical evidence supporting the alternative theory of the mind as embodied and social.

*Chapter Five*

# SYNERGIES AND THE INTERTWINED SELF

**THIS CHAPTER** involves some technical material, which I have suppressed in the main text. The discussion in this chapter is therefore, in a sense, incomplete. For those who want the mathematical details, I have included an appendix at the end of the book and pointers to specific sections in the appendix here in the main text.

Chapter 4 introduced radical embodied cognitive science as the scientific approach that goes with the ecological-enactive approach to the mind. The purpose of this chapter is to show how that approach can help us gather empirical evidence for the intertwined self. As in chapter 4, the technical details for the data analyses are in the appendix. We will see experimental evidence suggesting that we deeply and bodily intertwine with the world and with one another and that we accomplish complicated tasks together in virtue of that deep intertwining. Thus, the basis for our knowledge of one another is not simulation or inference. Although we might sometimes misunderstand one another, we have never been in Newton Haven.

## SYNERGIES

The scientific approach that we can use to gather evidence for the intertwined view of the self is a near relative of the work in coordination dynamics discussed in chapter 4. In a recent retrospective on coordination dynamics, Scott Kelso, the Kelso of the Haken-Kelso-Bunz (HKB) model, explains the connection between HKB and the scientific modeling we will be looking at in this chapter.

> Not only did HKB promote the mechanism of self-organization as an explanation of how new patterns of biological coordination could arise, persist and change, it brought to the fore the crucial roles of stability and variability. Up to then, variability was often seen as just "noise," something to be damped out or ignored. Now, experiments and theory were saying something else entirely, namely that variability is ubiquitous and essential for biological coordination. Among other aspects, variability allows the system to adapt and to change flexibly.[1]

Kelso's point is that the HKB model is not just an exercise in curve fitting, in which we take a theory-free approach to capturing regularities in data. It takes parts of living things, from neurons to limbs, to be self-organized collections of strongly coupled oscillators. Given what physicists have said about coupled oscillators, going as far back as Christiaan Huygens, HKB makes a series of predictions about the coordinated entities to which it is applied. One of these predictions is that there will be *critical fluctuations* as the coordinated system approaches a critical point, for example, just before it transitions from out-of-phase to in-phase coordination patterns. This prediction of critical fluctuation, which has been verified repeatedly in the cognitive sciences, is perhaps the clearest way in which the use of HKB allies the cognitive sciences with the sciences of complexity.

The key to gathering evidence for the intertwined self is to explain systems made up of individual humans, humans-plus-tools, and groups of humans as self-organizing, and therefore subject to the same kind of mathematical modeling that one applies to self-organizing systems in other sciences. A self-organizing system is one that exhibits regularities that arise without a plan or leader but emerge from the interactions of the parts of the system. The idea that human action is self-organizing began the recent resurgence of interest in dynamical systems models in psychology, which was discussed in chapter 4. Dynamical systems models work by assuming that thinking, experiencing, acting humans are self-organizing dynamical systems that comprise portions of their brains, bodies, and environments. Self-organizing systems have their organization without a plan or controller. They organize by using energy from their surroundings to create patterns, as Huygens's pendulum clocks do with gravity, HKB's wagging fingers do from the body's metabolism of sugars, and whirlpools in toilets do with the release of stored potential energy. Physicists who work on self-organization would call this behavior of pendula, fingers, and water molecules synergies.[2] A *synergy* is a collection of components that act as a temporary self-organizing unit and in which the components exhibit behavior that is constrained by being parts of the synergy (see figure 5.1, from Michael Riley, Michael Richardson, Kevin Shockley, and Veronica Ramenzoni).[3] The top portion of the figure depicts a set of components

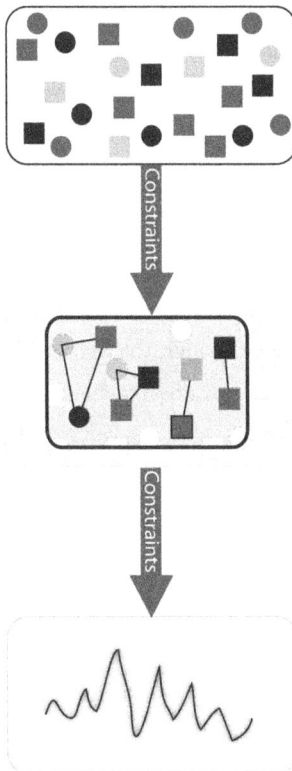

**FIGURE 5.1** A synergy. Used by permission of Michael Riley. CCBY.

as separate from each other, for example, a collection of water molecules in an unflushed toilet. The middle portion of the figure depicts constraints among some networks of those components so that the connected molecules behave collectively. The bottom portion depicts the synergy: constraints acting on the already constrained networks of molecules so that a coherent collective pattern of activity emerges.

Synergies might seem odd from a metaphysical or mechanistic point of view: a synergy is an aggregate entity that constrains the behavior of the smaller-scale components that make it up. Synergies thus imply a kind of downward or circular causation that strains our ordinary understanding of cause and effect. Nonetheless, there is nothing strange or magical about synergies—they occur routinely in nature and are well understood by physicists. The pattern in your flushing toilet is a synergy; convection rolls in boiling oil are synergies; bird flocks are synergies. In each of these cases, components form leaderless, temporary coalitions in which the behavior of the components (molecules, birds) is constrained by being part of the whole (the whirlpool, the flock) that they make up.[4]

One of the ways to identify synergies is that they are temporary coalitions that exhibit *interaction-dominant dynamics* and are therefore *interaction-dominant systems*.[5] These technical terms can be read quite literally: a system exhibits interaction-dominant dynamics when the interactions among the components dominate or override the dynamics that the components would exhibit separately. Systems with interaction-dominant dynamics (i.e., interaction-dominant systems) are genuinely unified systems. Over the last few decades, it has been demonstrated that many well-functioning physiological systems exhibit interaction-dominant dynamics. Examples include human heartbeats,[6] gait patterns,[7] and brain activity,[8] indicating that the chambers of the heart, the locomotory system, and parts of the human brain are interaction-dominant systems; that is, they are synergies.

Over the last few decades, synergies have made their way into the cognitive sciences, especially as practiced by radical embodied cognitive scientists. For example, there is mounting evidence that synergies are the chief mechanism by which brain areas work together to enable and control behavior—brain areas move into temporary coordinated patterns to enable animals to complete a task and become uncoordinated once the task is completed. For current purposes, the results from interaction dominance and physiology provide a way to gather evidence for the intertwined self: an individual organism and something outside its body (a tool, another organism) comprise a single, extended cognitive system if they are coupled to one another in the same way that the components of many well-functioning physiological systems are coupled to one another. That is, an organism and something outside its body comprise a single system if they collectively comprise a synergy with interaction-dominant dynamics.

## IDENTIFYING SYNERGIES

The key to finding empirical evidence for the intertwined self is finding evidence that human-tool and human-human systems are synergies. In the remainder of this chapter, I will briefly (and inadequately) set out the basic concepts and methodologies of a new research program.[9] In the next few sections, I show how, in particular instances, those concepts and methodologies have been put to use in gathering evidence for the intertwined self. This will again involve some technical details, which are described in more detail in the appendix. These methods have given rise to a new research program in the cognitive sciences and neurosciences concerning the character of fluctuations, a research program that promises to shed light on questions about cognition and the mind that previously seemed merely philosophical, like the intertwined self.

To repeat: a *synergy* or *interaction-dominant system* is a highly interconnected system, each of whose components alters the dynamics of many of the others to such an extent that the effects of the interactions are more powerful than the intrinsic dynamics of the components.[10] In an interaction-dominant system, inherent variability (i.e., random fluctuations or noise) in any individual component propagates through the system as a whole, altering the dynamics of the other components of the system. Because of the dense connections among the components of the systems, the alterations of the dynamics of these other components lead to alterations to the dynamics of the original component. That initial random fluctuation, in other words, will reverberate through the system for some time. So too, of course, would nonrandom changes in the dynamics of the components. This tendency for reverberations gives interaction-dominant systems what is referred to as "long memory."[11] In contrast, a system without this dense dynamical feedback, a *component-dominant system*, would not show this long memory. For example, imagine a computer program that controls a robotic arm. Although noise that creeps into the commands sent from the computer to the arm might lead to a weld that misses its intended mark by a few millimeters, that missed weld will not alter the behavior of the program when it is time for the next weld. Component-dominant systems do not have long memory. As is implied by the discussion of animal-environment coupling in chapter 4, in interaction-dominant systems, one cannot treat the components of the system in isolation: because of the widespread feedback in interaction-dominant systems, one cannot isolate components to determine exactly what their contribution is to particular behavior(s). And because the effects of interactions are more powerful than the intrinsic dynamics of the components, the behavior of the components in any particular interaction-dominant system is not predictable from the behavior of the components in isolation or from their behavior in some other interaction-dominant system. Interaction-dominant systems, in other words, are not *modular*. They are, in a deep way, *unified* in that the responsibility for system behavior is distributed across all the components.

We can see the distinction between component- and interaction-dominant systems clearly by looking at well-understood human physiological systems. Consider the human digestive system, depicted in figure 5.2, which is from a 1933 textbook.[12] Food enters the mouth and then passes through a series of structures (the stomach, the duodenum, the transverse colon, etc.), with each structure acting on the ingested food and then passing it on to the next structure. This is depicted on

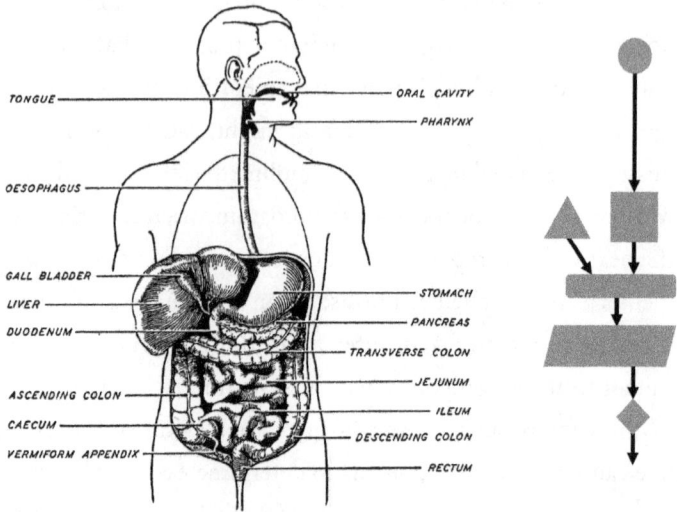

**FIGURE 5.2** *Left*: The human digestive system as depicted in a 1933 textbook. Public domain. *Right*: An abstract schematic rendering of the digestive system. Reproduced with permission from Luis H. Favela.

the right side of figure 5.2 as a box-and-arrow flowchart of the processing of food in a digestive system. It is a unidirectional path from inputs (from the mouth and the gall bladder) to output. Digestion is component dominant. Contrast this with the human heart. The heart is a collection of chambers and passageways between them, all of which interact with many other parts. Although blood does pass through the heart like food passes through the digestive system, the chambers and passageways do not merely act on blood and pass it along; they also constrain and control one another's behavior in real time. This is depicted in figure 5.3, also from 1933, which depicts the

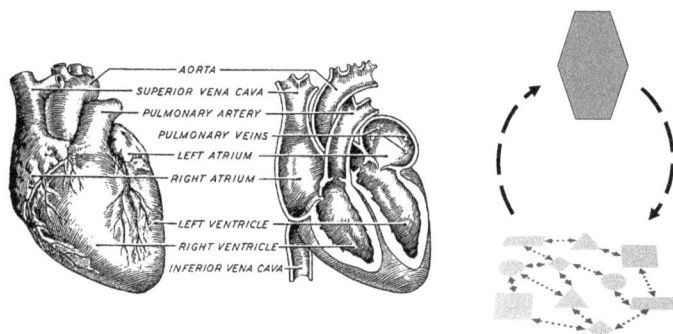

**FIGURE 5.3** *Left*: The human heart depicted in a 1933 textbook. Public domain. *Right*: An abstract schematic rendering of the human heart. Reproduced with permission from Luis H. Favela.

heart as an interaction-dominant system. In a healthy heart, the chambers and passageways are tightly interconnected so that minor fluctuations of the activity of the left ventricle affect the other chambers of the heart, which in turn affect the left ventricle so that the effects of this fluctuation will be seen in the heart for many beats into the future. In contrast, an unhealthy heart would have less strong connections among the parts so that the left ventricle fluctuation has no longer-term effect. It has been demonstrated that interaction-dominant dynamics is a strong indicator of heart health and that the loss of these long-term effects is a reliable predictor of cardiac arrest or other heart problems.[13]

As will be discussed in the rest of this chapter, this unity of interaction-dominant systems has been used to make strong claims about the nature of human cognition.[14] To use the unity of interaction-dominant systems to make these claims, it has to be argued that cognitive systems are, at least sometimes, inter-action dominant. This is where the critical fluctuations posited by the HKB model come in. Interaction-dominant systems self-organize, and some will exhibit self-organized criticality. *Self-organized criticality* is the tendency of an interaction-dominant system to maintain itself near critical boundaries so that small changes in the system or its environment can lead to large changes in overall system behavior. Self-organized criticality is a useful feature for behavioral control: a system near a critical boundary has built-in behavioral flexibility because it can easily cross the boundary to switch behaviors (e.g., go from out-of-phase to in-phase coordination patterns or from cantering to galloping). We see this all the time in professional sports: a National Football League quarterback running an option play skillfully maintains his behavior at the critical point between pitching the ball to a trailing halfback or running it himself.

In the next section, I describe a specific case of evidence that humans-plus-tools can be synergies, which is evidence that thinking, acting humans are not merely biological bodies. This is a key claim of the intertwined self.

## CASE STUDY 1: HUMAN-TOOL SYNERGY

Dobromir Dotov, Lin Nie, and I have used interaction dominance and synergies to test empirically a claim that can be derived from Martin Heidegger's phenomenological philosophy.[15] In particular, we wanted to see whether we could gather evidence supporting the transition Heidegger proposed from ready-to-hand to unready-to-hand interactions with tools. Heidegger argued that most of our experience of tools is unreflective, smooth coping with them. When we ride a bicycle competently, for example, we are not aware of the bicycle but of the street, the traffic conditions, and our route home. The bicycle itself recedes in our experience and becomes the thing with which we experience the road. In Heidegger's language, the bicycle is ready-to-hand, and we experience it as part of us, not different from our feet pushing the pedals. Sometimes, however, the derailleur doesn't respond right away or the brakes grab more forcefully than usual, and the bicycle becomes temporarily prominent in our experience; that is, we notice the bicycle. Heidegger would say that the bicycle has become unready-to-hand because our smooth use of it has been interrupted temporarily and it has become, for a short time, the object of our experience. (In a more permanent breakdown, like a flat tire, in which the bicycle becomes unusable, we experience it as present-at-hand.) To gather empirical evidence for this transition, we relied on dynamical modeling that uses the presence of $1/f$ noise or pink noise as a signal of sound physiological functioning, that is, as a signal of interaction dominance.[16] The connection between sound physiological functioning and $1/f$ noise allowed for a prediction related to Heidegger's transition: when interactions with a tool are ready-to-hand, the human-plus-tool should form a single interaction-dominant system, a synergy. This human-plus-tool system should exhibit $1/f$ noise (see $1/f$ noise in the appendix).

We showed this to be the case by having participants play a simple video game in which they used a mouse to control a cursor on a monitor. Their task was to move the cursor to "herd" moving objects to a circle at the center of the monitor. At some point during the trial, the connection between the mouse and cursor was temporarily disrupted so that movements of the cursor did not correspond to movements of the mouse before returning to normal. The primary variable of interest in the experiment was the accelerations of each participant's hand on the mouse. Accelerations are of special interest because they correspond to purposeful actions on the part of the participant, when they change the way their hand is moving. The experiment gave us a time series of accelerations for each participant. This time series was subjected to *detrended fluctuation analysis* (DFA). For details on DFA, see Human-Tool Synergy in the appendix.

The key finding was that, while participants were smoothly playing the video game, their hand movements exhibited $1/f$ noise and were part of a human-computer interaction-dominant system; when we temporarily disrupted performance, the $1/f$ noise temporarily disappeared. This is evidence that the mouse was experienced as ready-to-hand while it was working correctly and became unready-to-hand during the perturbation. That is, during smooth playing of the video game, the human-plus-mouse comprised a synergy. The fact that such a mundane experimental setup (i.e., using a computer mouse to control an object on a monitor) generated an extended cognitive system suggests that extended cognitive systems are common. Interacting with technology alters and extends your lived body, just as when a blind person uses a cane to cross the street or when you feel the street through your bicycle.

## CASE STUDY 2: INTERPERSONAL SYNERGIES

There have also been decades of research on *interpersonal synergies*. That pairs or collections of animals can form synergies should not be a surprise. For example, in bird flocks, no particular bird is in charge of the speed or direction of the flock, and the behavior of individual birds is determined by the behavior of the flock that they make up. The same sort of collective control is common in human interpersonal behavior. Remember the last time you danced or jogged with someone: you and your partner formed a temporary unit, in which each of you allowed your behavior to be constrained by the unit you comprised. The earliest experimental work on interpersonal synergies was done by Richard Schmidt, Claudia Carrello, and Michael Turvey.[17] They used the HKB model to show that while two people swing their limbs in synchrony, they are connected to one another in the same way that the limbs of a single individual are connected. That is, the interpersonal connection between two people is a temporary version of the intrapersonal connection among the parts of a single person. More recent work has shown similar interpersonal synergies arising when people engage in many interactions. For example, Michael Richardson and colleagues showed that participants seated next to one another on rocking chairs formed an interpersonal synergy, syncing their rocking when they collaborated on a task but not when they did not collaborate.[18]

Patrick Nalepka and colleagues (including Richardson and me) designed a two-person version of the herding video game described above.[19] In this version, two players stood on either side of a frosted glass tabletop onto which the video game was projected from below. Their task was to herd multiple objects ("sheep") to a circle in the center of the monitor ("the corral") using a motion-tracking puck that served both as their controller and as a motion sensor. They were successful and were allowed to leave the experiment early if they kept the sheep in the corral at the center of the screen for 70 percent of the last 45 seconds of eight consecutive trials. The participants were not allowed to speak to one another during or between trails. The basic finding is that the task is most easily solved when the participants stop paying attention to individual sheep and coordinate their behavior with one another. Successful pairs tended to engage in a search-and-rescue strategy at the beginning of trials, chasing down stray sheep and pushing them toward the corral. Once the sheep were in the corral, successful participants tended to engage in what we call coupled oscillatory containment, in which they each made coordinated semicircular movements around the outside of the corral on their side of the screen, either in phase with one another so that they would almost bump hands twice for each semicircle or out of phase with one another so that each participant was moving toward the point that their partner had just left. This finding was extremely robust. We were able to replace one of the participants with an artificial agent in a virtual environment, telling the human participant that they were playing with another human in another room. The results were the same. We were also able to implement the task in a virtual environment in which, rather than move a controller on a table with their hands, participants had to move their whole bodies around a three-dimensional space. Participants in the full-body movement succeeded by running back and forth, either in phase or out of phase with one another. (There was a difference, however. On the tabletop, in-phase coordination was more stable than out-of-phase coordination; when running back and forth, this was reversed.) Again, the key finding is that, to be successful participants had to coordinate their activities with one another rather than with the sheep. They formed an interpersonal synergy. (See the appendix.)

Instead of creating a novel video game, Ashley Walton and her colleagues (including Richardson and me)[20] studied the synergies that arise in jazz performance. She brought professional jazz piano players into the lab in pairs, suited them up in motion capture sensors, and had them improvise together over different backing tracks—the chord progression from a jazz standard; a single-tone drone; and an ostinato, a short, repeating musical phrase consisting of four ascending chords. We also had the musicians perform either in sight of one another or with a curtain drawn. We then dropped this condition because the coordination patterns between the musicians were the same whether they could see one another or not. We measured the musicians' hand movements and head movements, both up and down and side to side. At the beginning of the experiment, the musicians first performed two individual warm-up trials, where they improvised once over each backing track while their coparticipant waited outside the lab. Then together the pairs performed two sets of four two-minute improvised duets for a total of eight duets. Afterward, the musicians were interviewed separately about the performance while listening to their final few duets.

The coordination that emerged between the improvising musicians was analyzed using cross wavelet spectral analysis.[21] When Walton et al. had the pianists play along with a backing track together, they exhibited near perfect in-phase coordination with their hand and head movements. In contrast, when they improvised together, they had near perfect coordination in their up-and-down head movements, but the coordination in their melody-playing righthand movements disappeared; when there was coordination in hand movements, it was intermittent and tended to be out-of-phase. This might seem a puzzling finding because hand movements might seem more functionality relevant for musical production. However, head movements also play a large role in performance expressivity.[22] A compelling two-person improvisation demands that the players do not play the same melodies at the same time. The findings indicate that the interpersonal synergy that formed, and is seen in the head movements, is the backdrop for the improvisation. This coordination of head movements was unintentional and existed whether or not the musicians could see one another. This synergy is enabled because the musicians are expressing sensorimotor empathy—they are implicitly and nonreflectively coordinating their bodily activities to enable complex interaction. The musicians participate in an interpersonal synergy, which we can see in their coordinated head movements; this synergy is the foundation for their improvisation, which we can see in their melody playing. (See Jazz Performance in the appendix.)

## CASE STUDY 3: COMPLEXITY MATCHING

Humans routinely form interpersonal synergies, and doing so is beneficial. As we have just seen, interpersonal synergies are our basic bodily engagements with one another, the glue that makes possible more complex human interactions like having a conversation, playing a game, or making music. This basic behavioral phenomenon was discovered in 1999, when Tanya Chartrand and John Barge described the chameleon effect, which is the tendency for interacting humans to imitate one another's facial expressions, intonations, and gestures while engaged in conversation. Starting a few years later, cognitive scientists studying embodied social interaction focused on unintentional interpersonal synchrony. Kevin Shockley and colleagues showed that people engaged in conversation and joint problem solving unintentionally synchronized their postural sway and their gazes.[23] Richardson and colleagues showed that people who sat in rocking chairs while engaging in conversation and joint problem solving unintentionally synchronized their rocking.[24] Fabian Ramsmeyer and Wolfgang Tsacher showed that patients and therapists unintentionally synchronized their movements and that this synchrony improved patient outcomes.[25] Although the phrase wasn't yet available, these are studies of interpersonal synergy. The video game and jazz improvisation studies described above also focused on synchronization. However, not all interpersonal synergies are synchronic. Many tightly coordinated activities involved turn taking, for example. Consider two people dancing a tango. Or consider the coordination patterns in a large jazz orchestra, which might involve some synchrony but are hardly defined by it. Alexandra Paxton and Rick Dale, for example, showed that affiliative conversations are accompanied by more bodily synchrony than arguments, even though both still involve close coordination; arguing perturbs but does not disrupt interpersonal synergies.[26] More generally, to echo Ricardo Fusaroli, Joanna Rączaszek-Leonardi, and Kristian Tylén, all dialogues are interpersonal synergies.[27]

Drew Abney, Alexandra Paxton, Rick Dale, and Christopher Kello have shown that interpersonal synergies enable and structure dyadic conversation and problem solving by using *complexity matching*. Unlike the analyses of the last few sections, complexity matching makes sense of coordination over time without focusing on synchronization of the kind we saw in the head movements of the jazz performers. Complexity matching is also sensitive to coupling that results in turn taking and complementary dynamics, as are necessary in conversation so that only one person is talking at a time. Complexity matching demonstrates *nested temporal structure* in behavior. To compare, consider Duke Ellington's extended orchestral work *Black, Brown, and Beige*. It has an overall structure of six numbered parts; within each part, there are recurring motifs; within each motif, there are phrases played by different instruments; within each phrase, there are notes and movements of the musicians' bodies. This is the nested temporal structure of the piece. Complexity matching is the coordination of nested temporal structures between participants. Notice, however, that the complexity being matched is not a local feature of the behavior; it is the overall pattern of nested structure for the entire trial. The complexity scientist Bruce West and colleagues have shown that complexity matching is the most efficient means to transfer information between interacting systems.[28] Abney and colleagues looked for this sort of nested structure over the course of experimental trials. They measured bodily movements and vocalizations, but not for the content, function, or intent of the movements and vocalizations. Instead, they cared whether each participant was moving or not and speaking or not. (For more on complexity matching and the analyses Abney and colleagues used, see Complexity Matching in the appendix.)

The first study Abney and colleagues did on complexity matching was a reanalysis of the data from Paxton and Dale's demonstration that argument disrupts interpersonal synchrony.[29] They found that, in affiliative conversation, participants coordinated their patterns of nested temporal structure more than in argumentative conversations; that is, they matched complexity more in affiliative conversation than in arguments. A second complexity matching study by Abney, Paxton, Dale, and Kello involved a building task. They had pairs of participants collaboratively build towers from marshmallows and uncooked spaghetti on a tabletop. One participant could touch only the spaghetti and the other could touch only the marshmallows. They were told to build the tallest tower they could in fifteen minutes. This is admittedly a strange task, but it is open-ended and relatively unconstrained; it also requires cooperation and natural conversation from the participants. They found that vocal complexity matching predicts success at the task. Pairs who matched their vocal complexity built taller towers. These findings make sense when understood in terms of the work by West and colleagues. Coordination in conversation has long been thought to depend on the establishment of a common ground of shared facts and attitudes.[30] The enhancement of information transfer in complexity matching is the most efficient way to establish that common ground; less common ground is required for arguments.

These studies show that individual humans become parts of interpersonal synergies by matching the complexity of their activity with one another. This complexity matching is a feature of our basic bodily interactions. In conversation, we match one another's complexity, even though complexity is a feature of longer timescales and is not consciously accessible. That is, I cannot consciously perceive the complexity of my own behavior or the complexity of yours, but when we socially interact, I manage to match my complexity to yours. This suggests that we do not form interpersonal synergies by theorizing about or simulating one another. As Shaun Gallagher has it in his interaction theory, basic bodily interaction is not a matter of thinking about one another. It is a matter of sensorimotor empathy, our skillful, implicit basic bodily being with one another.

## CONCLUSION

The purpose of this chapter has been to accrue evidence for the intertwined self. I have provided empirical evidence that we routinely form interpersonal synergies with one another and that these synergies do not depend on having a theory of mind or the ability to simulate other minds. We stick to one another when we interact, forming multiperson systems that have the same dynamic signature as well-functioning biological systems and enable efficient information transfer between the components. Our real-time social connections are interaction-dominant, like the human heart, not component-dominant, like an iPad or the human digestive system. These interpersonal synergies are a straightforward demonstration that self-other blurring is common indeed.

Interpersonal synergies are not consistent with the Cartesian assumptions that define the modern theory of the mind. They imply that the human mind is not a hidden, inner machine separable from the body, from action, from other humans. Interpersonal synergies are not a matter of forming speculative models of one another's hidden thoughts. When two humans engage in a conversation or cooperative video game, they are not physically connected in the way that water molecules are. This scientific research on synergies reinforces the claims made by phenomenologists and radical embodied cognitive scientists that perceiving, experiencing, thinking, and the like, are things that we do, not things that happen in our brains. We do them by forming tight connections to the world around us, not by building internal models of the world around us. The scientific research on interpersonal synergies suggests that perceiving, thinking, experiencing, and the like, are things we do regularly in concert with other people, forming connections that we temporarily structure and that temporarily structure us.

# PART III

*Now philosophy has become mundane, and the most striking proof of this is that philosophical consciousness itself has been drawn into the torment of the struggle, not only externally but also internally. But, if constructing the future and settling everything for all times are not our affair, it is all the more clear what we have to accomplish at present: I am referring to* ruthless criticism *of all that exists, ruthless both in the sense of not being afraid of the results it arrives at and in the sense of being just as little afraid of conflict with the powers that be.*

Marx to Ruge, September 1843, *Marx-Engels Collected Works*

*Chapter Six*

# SOCIAL ONTOLOGY, REPRESENTATION HUNGER, AND THE INTERTWINED SELF

**IN CHAPTERS 4 AND 5,** I set out a scientific approach that combines the ecological and enactive approaches with nonlinear dynamical modeling and used it to present evidence supporting sensorimotor empathy and the intertwined self. Using work in nonlinear dynamical cognitive science, we saw that self-organization occurs at multiple scales in human action and physiology, from heart beats to brain activity, to bodily movement. This suggests that hearts, brains, and movements are synergies, systems composed of multiple interacting parts constrained to act temporarily as a unit. We also saw that human-tool and interpersonal synergies are common and have the very same dynamics as these individual processes, suggesting that multiperson systems are just as deserving of status as real entities as hearts, brains, and bodies. This is evidence for the intertwined self: human-tool and social dynamical systems have biological integrity similar to individual human bodies. We are in the world and with others.

We know that others are not blanks (as in the film *The World's End*) by interacting with them and forming interpersonal synergies. What we call a self is partly constituted by these interpersonal synergies. The minimal self is social. If the commonly understood nesting of self as a component of the mind and the mind as a component of the person is accepted, this implies that minds and persons are necessarily in the world and with others.

In part III of this book, I will attempt to draw some lessons from the foregoing. I will relate the story to issues in philosophy of mind and cognitive science, the location of cognition, the question of group cognition, and the so-called representation hunger problem for theories of cognition that don't assume that it involves representation. Then I will venture beyond theorizing about the mind per se to related issues in ethics and politics.

## GROUP MINDS, DISTRIBUTED COGNITIVE SYSTEMS, AND JOINT ACTION

We talk routinely about groups of humans as if they, the groups themselves, have thoughts and experiences: the mood of the crowd, the desires of a corporation, the beliefs of a political party, even the goals of the ant colony. One of the sites of good philosophical work this century has been the debates about whether to take such talk as literal or figurative.[1] It might seem obvious that the intertwined self would requires us to take sides in these debates. After all, I have argued that even the minimal self is social. It does require us to take sides, but not necessarily in the way one might think. First, we must define some terms.

Much of the early talk about the thoughts and experiences of groups referred to so-called group minds. In 1921, while trying to make sense of the horrors of World War I, which had just ended, William McDougall wrote, "I have argued that we may properly speak of a group mind, and that each of the most developed nations of the present time may be regarded as in the process of developing a group mind."[2] The question throughout the twentieth century was whether what seemed to be group minds were entities in their own right[3] or mere summations of the minds of the individuals who made up the group.[4] As a way to avoid some of the harder metaphysical questions concerning consciousness and the normative requirements of common-sense psychological notions like belief or intention, the discussion in the twenty-first century turned mostly to what are called distributed cognitive systems.[5]

Distributed cognitive systems are beautifully exemplified by Edwin Hutchins's[6] studies of navigation abord US naval ships. Hutchins describes the participants in on-ship navigation in bays and harbors as a multiperson, multitool computational process. The fix cycle, which is done every three minutes in bays and harbors begins with two people on deck, using tools called alidades to find the angular difference from magnetic north and two different objects; each person relays a number via telephone to another person who records those numbers along with the time in a ledger. A fourth person uses another tool, a hoey, to draw lines on a map an appropriate number of degrees from the representation of the object. If the ship were not moving, it would be at the point on the map where those two lines intersect. Because the ship is moving, the process is repeated twice, yielding three points on the map that form a triangle. The ship is in this triangle. Locating the ship in this way is accomplished by four humans interacting with tools and with one another. The key, in Hutchins's description, is that this works, even though none of the individual humans might know how to complete the whole task or what role their activities play in the task; the task and the knowledge how to accomplish it are distributed across the whole distributed cognitive system.

In one widely discussed article, Georg Theiner, Colin Allen, and Rob Goldstone[7] argue for distributed cognitive systems as genuine cognitive systems by using what they call the social parity principle. The social parity principle is derived from the parity principle described by Andy Clark and David Chalmers in their famous article "The Extended Mind": "If, as we confront some task, a part of the world functions as a process which, *were it done in the head*, we would have no hesitation in recognizing as part of the cognitive process, then that part of the world is part of the cognitive process."[8] Clark and Chalmers argue that this principle suggests that things like eyeglasses, notebooks, and smartphones are genuine parts of extended cognitive processes. Theiner, Allen, and Goldstone alter this slightly for their social parity principle: "If, in confronting some task, a group collectively functions in a process which, were it done in the head, would be accepted as a cognitive process, then that group is performing that cognitive process." Theiner and colleagues then proceed to pile on examples of things done by groups that would be considered cognitive had they occurred inside a brain. I will describe just one here: transactive memory systems, which were introduced by psychologist Daniel

Wegner[9] to discuss how memory is organized in established groups of individuals, like families, couples, and teams of work colleagues. It is well established that participants in transactive memory systems remember a particular subset of things, know who knows what, and share a set of procedures that they use to encode new memories, recall memories of the group, allocate the storage of memories, share stored memories with one another, and elaborate memories. Notice that this list of features of transactive memory systems is exactly what one would expect in a computer or an individual recalling something. If we were to learn that inside each of us there were brain areas where different kinds of memories would be stored and another set of brain areas that would instantiate a series of procedures for routing memories to the appropriate brain areas for storage, retrieving memories from where they are stored, and combining them with other memories, we would not hesitate to call these brain areas and their activities the individual's memory. By the social parity principle, the activities of the group of people that compose a transactive memory system are genuine memory and the group collectively makes up a genuine cognitive system.

Hutchins makes the point that these distributive cognitive systems are the best evidence we have for the computational theory of mind. When we watch a crew piloting a large ship, we can actually observe the computations as they happen. When the sailor on deck uses the alidade to report a number to the bridge, they are visibly doing an analog-to-digital conversion: the information in the light is continuous, that is, analog, and the number they speak into the phone is digital. There are similar analog-to-digital and digital-to-analog conversions that occur throughout the fix cycle. These are computations and they are straightforwardly observable; we can see them happening. This is in sharp contrast to the attempts to observe computations inside the brains of humans. Whether a pattern of activity in a brain area represents something or computes a function is not detectable, even with our best imaging equipment. The brain activity that might represent, say, a grandmother's face, is not observably different from brain activity that might do a digital-to-analog conversion. This means that speculations about what a brain area represents or what function it computes are always that: speculations.[10] Distributed cognition is often straightforwardly, observably computational.

The "groups are real entities" versus "groups are mere summations" debate does not go away just because we talk about distributed cognition instead of group mind.[11] Kirk Ludwig, for example, argues for the mere-summations side in an article-length response to Theiner et al. and in a two-volume work.[12] He writes

...what a group cognitive level process of the sort that we are interested in requires is that there be group level intentional states, a group level thinker, or cognizer, a group level possessor of representations of the task, a group level desire and intention to pursue it, and group level beliefs about how to do it, and, hence, a group level agent. Nothing follows, as we have seen, about there being a group level agent from the fact that the group solves the problem, because all this comes to is saying that each member of the group intentionally contributes to its solution.[13]

This argument suggests that the transactive memory systems that Theiner et al. discuss are not themselves cognitive systems because the systems do not themselves have thoughts, representations, desires, and beliefs. Ludwig thinks that only the individual agents who compose things like transactive memory systems do, and the group exists because each individual purposefully tries to solve a common problem. The individuals are the ones with thoughts, representations, desires, and beliefs, and those exist inside the individuals. We can argue about whether this works as an argument against transactive memory systems as group cognitive systems, but it definitely does not work as an argument against the sailors engaged in the fix cycle comprising a genuine group cognitive system. The main reason for this is that the participants in the fix cycle are not typically trying to solve a common problem. As Hutchins points out, not everyone in the system even needs to know what the problem to be solved by the system is and not everyone has to be intentionally contributing to solving it. In the case of the fix cycle, Hutchins points out, the problem would get solved even if everyone was not aware of the problem as a whole.

Even more clear, however, is that this whole debate simply assumes that the modern theory of the mind is correct. Consider the parity principles described above:

> If, as we confront some task, a part of the world functions (a group collectively functions) as a process which, *were it done in the head*, we would have no hesitation in recognizing as part of the cognitive process, then that part of the world is part of the cognitive process (that group is performing that cognitive process).

These principles make clear that Clark, Chalmers, Theiner, Allen, Goldstone, and Ludwig all agree that the basic, incontrovertible cases of cognition are those that occur in the head; that is, they all agree that most thinking and experiencing takes place in a hidden, inner realm, invisible to outsiders. The main aim of this book so far has been to argue that this understanding of the mind is mistaken. Suppose it is mistaken. Where does this leave us in the arguments over group minds and distributed cognition?

The first thing is that if humans are basically embodied and social, as I have been arguing, we should expect group cognition to be the norm.[14] If even the minimal self is intertwined with others, we should expect most of the rest of our thinking and experience to be shared with others. Given what I have discussed in the previous chapters, this should not be a surprising claim. It echoes a point made by Edward Baggs[15] in the context of the ecological approach. Recall that in the ecological approach, the primary entities that are experienced are affordances, or opportunities for behavior. Baggs argues that the debates over what makes some of the affordances social is misplaced because all affordances are social. Martin Heidegger claimed that being is first and foremost being-with. Maurice Merleau-Ponty claimed that we are in one another. These all suggest that groups of humans are genuine cognitive systems because they are foundational in being individual humans. In chapter 5, we also saw empirical evidence that humans-plus-tools and groups of humans have their own emergent dynamics and that these dynamics constrain the activities of the individual humans that make them up.

From the point of view of the intertwined self, group cognitive systems should be foundational and not something we might potentially derive from individual cognitive systems. In fact, this problematic is inverted by the intertwined self: given that even the minimal self is social, the existence of individual cognitive systems is what requires explaining. We will discuss this in chapter 7.

There is a deeper issue related to Ludwig's reply to Theiner et al. For Ludwig, being a cognitive system requires representations of the cognitive task being carried out. But for phenomenologists and radical embodied cognitive scientists, this is not a requirement for being a cognitive system; all these scholars are skeptical that thinking, perceiving, experiencing, and the like, are to be solely understood in terms of internal representations of an external world. Also notice that none of the scientific explanations of human-tool and human-human systems discussed in chapter 5 refer to representations. Proponents of the intertwined self must reject the characterization of cognitive system that Ludwig proposes. Cognition is not defined in terms of internal representations of the environment or of tasks. This requires rejecting the parity principles as originally written, which simply assume that the base case of cognition is hidden and inner. Rejecting that assumption actually makes human-tool and human-human cognitive systems more plausible. Believing that cognition necessarily involves representations invites the following response to Clark and Chalmers's claim that smartphones are part of cognitive systems: it is not the smartphone itself that is part of the cognitive system; it is only the human user's internal perceptual representation of the smartphone that is genuinely part of the cognitive system. Cognitive scientists Natalie Sebanz and Günther Knoblich argue that joint action—coordinated, goal-oriented engagement of two or more individuals completing a task—requires that each participant in the action have internal representations of the intentions and likely future actions of the others.[16] As with the smartphone users, these hidden, inner representations of partners are part of each individual human cognitive system that participates in the joint action; there is no genuine group cognitive system.

In rejecting the modern theory of mind, we must replace the "were it to take place in the head" part of the parity assumptions with "were it done by an individual biological organism." We are left instead with something much more commonsensical.

> If, as we confront some task, a part of the world functions as a part of a process which, were it done by an individual biological organism, we would have no hesitation in recognizing as part of the cognitive process, then that part of the world is part of the cognitive process.

and

> If, in confronting some task, a group collectively functions in a process which, were it done by an individual biological organism, would be accepted as a cognitive process, then that group is performing that cognitive process.

This substitution removes the assumption of the hidden, inner mind and allows genuine human-tool and group cognitive systems. Sometimes you remember telephone numbers; sometimes you use your smartphone to remember telephone numbers. Sometimes you remember all or parts of a story; sometimes, in transactive memory, you use a long-term partner to remember parts of it. Deborah Tollefsen, Rick Dale, and Alexandra Paxton have made a related point in their discussion of what they call alignment systems.[17] An alignment system is a loosely interconnected set of cognitive processes that facilitate social interactions. Alignment systems are dynamic, multiscale, and multicomponent systems that are responsive to our intentions to engage with others, as in conversation or improvising music. Coupled alignment systems can also give rise to such shared activities. Humans engaged in social tasks form interpersonal synergies and are coupled to one another at multiple, interacting spatial and temporal scales. There are some tasks that we are more comfortable calling cognitive than others. In Walton's jazz improvisers, the bodily coordination seems less straightforwardly cognitive than the interacting melody playing. But Tollefsen, Dale, and Paxton argue that what we see instead here is coupled cognitive systems at multiple scales so that the bodily coordination is no less genuinely cognitive than the playing of the melody. This might seem too loose a requirement on being (part of) a cognitive system, but it is absolutely required if we want to rule out blanks. If you think that people need any particular process happening invisibly inside them for them to count as thinkers, you will never escape from Newton Haven.

## THE BRAIN-IN-THE-ASS HYPOTHESIS

There is something of a tradition to mention the song "Where Is My Mind?" by the Pixies when discussing understandings of cognition as extended or social. Andy Clark, for example, uses the title of the song as the opening of a *New York Times* essay about the extended mind.[18] It answers the question about the location of the mind, or at least Pixies singer Black Francis's mind: swimming "way out on the water," which is to say not in his brain or even his head. Fred Cummins, in a blog post from 2015,[19] also leads with the Pixies song, although the reference does not make it to the related discussion in his groundbreaking book *The Ground from Which We Speak*.[20] In his discussion, Cummins offers an alternative to the infamous "brain-in-the-vat" thought experiment, which originally suggested that, for all you know, you could be just a brain, floating in a vat that is being maintained and fed inputs by an evil genius scientist. This thought experiment simply assumes the modern theory of the mind that I have been arguing against. It is also a very bad thought experiment, for what it is worth: Diego Cosmelli and Evan Thompson decisively refute it just by wondering what it would really take to keep a brain living and having experiences. They point out the following three features of brain activity:

1. It is largely generated endogenously and spontaneously.
2. It requires massive resources and regulatory processes from the rest of the body.

3. It plays crucial roles in metabolic and regulatory processes that keep the organism alive, and these processes require a constant influx and outflux of informational and chemical resources.

Together, these three points suggest that the only way a brain in a vat is possible is if the vat is equivalent to a biological body interacting with an actual environment. "In short, the so-called vat would be no vat at all, but rather some kind of autonomous embodied agent."[21]

Unlike the brain in the vat, Cummins's alternative thought experiment is not intended to inspire epistemological dread or terrible movies. Cummins is interrogating the intuition that most of us have about the location of our experiences. Doesn't it feel like your experience happens behind your eyes and between your ears, that is, in your brain? Even Clark, with his view of the mind extended into the world in various ways, thinks that experience still happens in the brain. In attempting to dislodge this intuition, Cummins proposes the "brain-in-the-ass" hypothesis. Suppose, Cummins asks, that your brain is not removed from your body and placed in a jar but simply sent to the other end of your spinal cord. This would take some "relatively trivial" rewiring and some anatomical changes to replicate the protection that our skulls offer; it would likely also take some adjustments to neuronal propagation velocities and things like that. The brain in the ass is still implausible, but certainly no more implausible than the brain in the vat. In fact, this brain-in-the-ass scenario is not subject to the same concerns Cosmelli and

Thompson bring up: the brain in the ass is in the body and plays a role in how that body engages with the environment. It is just in a different location. Suppose we could move the brain in this way. Where would it feel like your experiences are happening? Your eyes would still be in the same place, connected to (longer, but faster) optic nerves, and moved by the same muscles and brain regions; your ears would still be in the same place, and vibrations would affect the same hairs in your ear; your nose and mouth also would not move. Cummins suggests, and I agree, that even if your brain were at the other end of your spinal cord, you would think that your experiences happen in your head because that is where the primary organs of most of your senses would still be. Cummins points out that even your inner voice, with which you engage in conscious thought, would still feel like it happens in your head given the way your typically hear yourself talk out loud: our voices stimulate the hairs in our ears, but they also vibrate the entire skull, and most of our hearing of our own voice comes via bone conductance. But it would seem less obvious to you if your sensory experiences and inner voice happen in your brain because your brain is somewhere else. As Cummins puts it, "Shift the brain, and nothing happens. Just be careful sitting down."[22]

The intertwined self implies that experiences and thought are not hidden, inner neurological events but occur in the world, with technology and with other people. This implies that the answer to the Pixies' question is not "In my brain," but rather spread throughout your body and the tools you use and the people you interact with. It is also not swimming far from shore as the Pixies suggest, unless that is what you are doing.

## PUBLIC, OUTER, COMPLEX COGNITION

One common response to the sorts of ideas that we have been discussing is to point out that, although things like alignment systems might be part of the explanation about some of our cognition, there are some types of our cognition that could never be explained this way. One way this is put, originally by Andy Clark and Josefa Toribio, is to say that some kinds of thinking are representation hungry in that it is hard to imagine explaining them without referring to operations on internal representations of the world. Alignment systems and interpersonal synergies are natural explanations of basic bodily and social coordination dynamics, this objection tends to go, but they will never explain real thinking. Real thinking, Clark and Toribio say, "involves reasoning about absent, non-existent, or counterfactual states of affairs."[23] Examples include imagining what things would be like if pigs could fly or if all of the members of the UN General Assembly were women. Accomplishing tasks like these, it is argued, seems to require hidden, inner processes. The argumentative burden for those of us who do not think that internal representations are among the requirements for real thinking is to explain how it could be explained without them.

Since Clark and Toribio made this objection, there have been a series of responses pointing to cases in which hidden, inner processes are not required to explain how humans engage in real thinking like this. For example, Damian Stephen and Jay Dixon explained problem-solving insight as dynamical and not requiring representations.[24] Iris van Rooij, Raoul Bongers, and Pim Haselager explained imagining future actions without

internally representing those actions.[25] Perceiving any affordance is a matter of perceiving possible future actions, which
are absent, non-existent, and counterfactual. When I perceive
that a table toward the side at a concert venue affords sitting for
me and my friends, I am perceiving a possible future, something
that hasn't happened and might never happen. As we saw in
chapter 4, proponents of the ecological approach never explain
perception of affordances in terms of internal representations.
One thing to draw from this discussion is that we shouldn't
put much stock in the inabilities of even very smart people like
Clark and Toribio to imagine explanations. We should never
be convinced by what we might call argument from lack of
imagination. I can't imagine how so-called nonfungible tokens
could be anything but a scam to sell cryptocurrency, but my
lack of imagination shouldn't carry much argumentative weight.
Someday, someone might explain their real value.

Clark and Toribio's argument is a symptom of a widespread assumption in the cognitive sciences that the point of
real thinking, like imagination or creativity, is to pull us away
from our engagement with the world. For example, Peter
Langland-Hassan identifies four distinct types of imagination:
imagistic imagining, conditional reasoning, pretense, and con-
suming fiction. He explains each type of imagination not as its
own psychological faculty but as a combination of many other
faculties with which we internally represent the world, such as
our abilities to form beliefs, desires, and judgments.[26] In one
of the most influential works on the cognitive science of creativity, Margaret Boden distinguishes three forms of creativity.[27] The first involves generating unfamiliar combinations of

familiar ideas, as when one thinks of poetic imagery or a new melody; the second is the exploration of conceptual spaces, as is done by scientists and philosophers; the third is realizing the limitations of conceptual spaces and transforming them to surpass those limits. For current purposes, the key uniting factor in these three types of creativity is that they all occur in the realm of hidden, inner mental representation. However, there is good reason to think that these cases of real thinking do not involve a psychologically pulling away from the world. Perceiving affordances, for example, is a matter of imagining possible futures, but perceiving affordances is in no way a kind of recession into a private mental realm.

There is a long-standing tradition in philosophy, anthropology, and psychology that sees imagination and creativity as kinds of engagement with the world and with other humans. Emilien Dereclenne traces this understanding of imagination all the way back to the Renaissance philosopher Giordano Bruno.[28] We will start in the more recent past, with Tim Ingold.[29] Following Gilbert Simondon and Gilles Deleuze (more on Deleuze in chapter 8), Ingold critiques what he calls the hylomorphic model behind most understandings of creativity. According to the hylomorphic model, creativity is the generation of novel ideas, and creative activity is the imposition of these ideas onto matter. This is exemplified in views like Langland-Hassan's and Boden's, where the imagination and creativity end before any actual making of artworks or artifacts occurs. In contrast, Ingold proposes a view of imagination and creativity in which they exist not in an inner realm but in the act of making. Ingold aims to overthrow the hylomorphic view and replace it with a view that assigns primacy to the material processes of making. Ingold provides vivid accounts of basket weaving, architecture, and cooking to illustrate how different practices of making involve "not so much imposing form on matter as bringing together diverse materials and combining or redirecting their flow in the anticipation of what might emerge."[30] In making, the imagination and creativity do not occur in the mind prior to interacting with materials but rather in the interaction with the materials. For Ingold, "imagination is not a mental capacity that permits the spontaneous generation of ideas, but rather a way of living creatively in a world that is itself crescent, always in formation."[31] Lambros Malafouris reaches similar conclusions from his studies of traditional Greek potters.[32] The process of forming a clay pot, by spinning the wheel and molding the wet clay with one's hands, is a nonlinear interaction between a skilled craftsperson and raw materials, both of which have agency and together determine the form of the resulting pot. Malafouris's material engagement theory explicitly rejects the hylomorphic conception of imagination and creativity and argues that the final form of a clay pot is not the externalization of something from the mind of the potter.

> Material-engagement theory proposes that forming and thinking are inseparable. The form that we see emerging is not the product of externalization. On the contrary, material form is folded into the mental. The potter's projections and anticipations inhabit clay. Forms do not travel from mind to matter. Form making is more of a gift exchange than an imposition.[33]

In material engagement theory, the materials with which we interact are inseparably intertwined with our creative activities. "The claim is that things actively participate in human cognitive life or that human thinking is better described as *thinging*. We think with and through things, not simply about things."[34] The creativity and imagination of any maker, in this view, occurs in thinging.[35]

This understanding of creativity and imagination as public and outer is hardly the exclusive domain of anthropologists. Rob Withagen and John van der Kamp put Ingold's ideas in the context of the ecological approach in psychology.[36] Withagen and van der Kamp argue that, because action is mostly unplanned and improvisatory, creativity is found in action itself, especially when action uncovers novel affordances. In cooking or forming a pot on the wheel, we use the affordances of the raw materials, and our interactions with them to create something that is almost always unique. That this Ingold-Gibson mashup is possible is in a way unsurprising: for a portion of his career, Ingold thought of himself as a Gibsonian.[37] Vladimir Glăveanu, Wendy Ross, and Frédéric Vallée-Tourangeau have done experimental work on creativity and imagination that parallels Malafouris's work on thinging. Glăveanu understands creativity as a form of action in and on the world, embedded in a rich sociocultural context. Ross and Vallée-Tourangeau discuss creativity and imagination in terms of what they call kinenoesis, in which knowledge about how to act and to solve problems is built into movements.[38]

For all these thinkers, imagination and creativity, which are undisputed exemplars of real thinking, are neither hidden nor inner. This set of examples differs from those in two recent manifestos, one by Michael Spivey, the other by Mark Dingenmanse and colleagues, but they make the same point.[39] The progress that has occurred in the cognitive sciences since the inception of the field in the 1950s has been a slow and steady march toward taking interaction—with the environment, with technology, with others—as the key to explaining thinking. Cognitive scientists are slowly moving away from the modern theory of the mind as hidden and inner.

*Chapter Seven*

# THE PRAGMATIST TRADITION AND INNER SPEECH

## INTRODUCTION

In this chapter and the next, I will put the ideas discussed so far in this book into context: historical, intellectual, and political. In this chapter, I will discuss further twentieth-century precursors to the intertwined self. I will use these ideas to examine more closely how one can explain another type of real thinking, in this case inner speech, without assuming that the mind is, in general, hidden and inner. This is important because although I have been arguing that we do directly perceive that other humans are not blanks (as in the film *The World's End*), we do not thereby know all their thoughts. Our inner monologues, despite their social origin, can be private.

## THE EARLY TWENTIETH CENTURY:
## PRAGMATISM AND INTERACTIONISM

Pragmatism as a philosophical position developed from a philosophical conversation club, called the Metaphysical Club in Cambridge, Massachusetts, in the late nineteenth century. Among the members of the club were William James, Charles Saunders Peirce, and Oliver Wendell Holmes.[1] When he first used the term "pragmatism" in print in 1898, James credited the idea to Peirce, likely as a favor to his then-unemployed friend. Peirce rejected the attribution and the philosophical position that James laid out and renamed his own philosophical position "pragmaticism." Jamesian pragmatism became very influential. The idea is essentially Darwinism about the mind. Just as human lungs are adaptations to our long-standing oxygen-rich environment, my perceptions, thoughts, and experiences are short-term adaptations for the situation I find myself in and are inseparable from the actions I undertake; that is, the meaning of my thoughts is a function of the actions they lead me to engage in. To be true, my thoughts must lead me to appropriate action, given my situation. "My situation" in these sentences should be viewed expansively: not just the immediate environment here and now but as extended in time and space, and including my developmental, social, and cultural settings. The resulting picture of the mind is as a set of habits and abilities, inseparable from the body and material and social contexts, for achieving an organism's ends. This set of ideas was extremely influential in psychology, philosophy, and especially education in the early twentieth century.[2]

The key player in pragmatism's influence in education was John Dewey. Dewey applied pragmatism to education when he founded the Laboratory School at the University of Chicago in 1896. The school was a laboratory both for Dewey and for the students who attended. For Dewey, it was a place to test his psychological and philosophical ideas; the students learned by doing self-directed experiments, with their teachers serving as facilitators and guiding them only loosely. (This is the beginning of student-centered education.) Among the things that Dewey learned in his observations of children at the school was that learning is intensely social. Children learn by spontaneous doing, always accompanied by a lot of talking and listening.[3] Later, in one of his most important books *Experience and Nature*, Dewey makes this sociality central to his understanding of the mind. "Personality, selfhood, subjectivity, are eventual functions that emerge with complexly organized interactions, organic and social."[4] Dewey's colleague at the University of Chicago, George Herbert Mead, took the ideas Dewey developed by studying children and applied them to settlement houses. Settlement houses were established in response to the concentrations of urban poverty that followed the industrial revolution. In settlement houses, middle-class people would relocate to poor neighborhoods to work with local residents on programs aimed at alleviating the suffering that comes with poverty. The most famous settlement house was Chicago's Hull House, which was run by Dewey's and Mead's friend Jane Addams; Mead served for a time as the treasurer at Hull House. In Mead's view, settlement houses were laboratories in the same sense that Dewey's school was. Mead contrasts settlements with both scholarly work and religious missionary work. Unlike missionaries, settlers moved to a neighborhood without pre-conceived ideas about what residents needed but with plans to learn from them. Unlike scholars, settlers do not aim merely to observe their neighbors but to facilitate attempts to alleviate suffering. More important, and unlike missionaries, settlements are "not primarily engaged in fighting evils but in finding out what the evils are; not in enforcing preformed moral judgments, but in forming new moral judgments."[5] In the pragmatist tradition, a settlement "is discovering proper lines of conduct, not primarily facts. The settlement is practical in its attitude, but inquiring and scientific in its method."[6]

From his work in settlements, Mead developed an influential understanding of the self as necessarily embodied and social. The best encapsulation of Mead's views on the self are found in a posthumously published book *Mind, Self, and Society*. Mead ties the emergence of the individual mind closely to the development of linguistic abilities:

> Out of language emerges the field of mind. . . . It is absurd to look at the mind simply from the standpoint of the individual human organism; for, although it has its focus there, it is essentially a social phenomenon; even its biological functions are primarily social. The subjective experience of the individual must be brought into relation with the natural, sociobiological activities of the brain in order to render an acceptable account of mind possible at all; and this can be done only if the social nature of mind is recognized. The meagerness of individual experience in isolation from the processes of social

experience—in isolation from its social environment—should, moreover, be apparent. We must regard mind, then, as arising and developing within the social process, within the empirical matrix of social interactions. We must, that is, get an inner individual experience from the standpoint of social acts which include the experiences of separate individuals in a social context wherein those individuals interact. The processes of experience which the human brain makes possible are made possible only for a group of interacting individuals: only for individual organisms which are members of a society; not for the individual organism in isolation from other individual organisms.[7]

For Mead, the mind is necessarily social; there are no experiences outside a social context.[8]

At the same time, but far from the settlement houses and the capitalist context in which pragmatism emerges, the Soviet psychologist Lev Vygotsky developed an understanding of the mind and self very similar to Mead's.[9] Vygotsky thought that individual mental lives resulted from the internalization of cultural tools during development. Like Dewey, Vygotsky understood learning as an intensely social process. This is exemplified in what he calls the zone of proximal development, which is the difference between what a child can do on their own and what they can do with the help of adults or older children. As infants, children can do essentially nothing on their own and can do things only with the help of a caregiver. That is, all their early problem-solving skills are social. For Vygotsky, this never goes away. Young children always learn with the help of others, including by speaking and listening. Only through speaking with and listening to others can children develop an inner voice so that they can talk quietly to themselves. That is, what we typically think of as an individual thinker appears after years of social interaction. As for Mead, for Vygotsky the mind is necessarily social.[10]

The parallels between Vygotsky and Mead are striking, especially because they never interacted. The isolation of Soviet science from work in Europe and the Americas did not happen until the Cold War, after both Mead and Vygotsky were dead. Dewey did travel to the Soviet Union and was very impressed by it and other "revolutionary" societies,[11] but there is no record of Dewey meeting Vygotsky. Vygotsky was aware of pragmatism and of William James's *Principles of Psychology*,[12] but not of Mead; Mead was not aware of Vygotsky's work, none of which appeared in English until the 1960s. Why then are their ideas so similar? Some historians think that the similarities are nothing more than a curious coincidence. Others point to the influence that G. W. F. Hegel had on both: directly and via Karl Marx for Vygotsky; directly and via Dewey and Josiah Royce for Mead.[13] However they got there, both thinkers developed near relatives of what I have been calling the intertwined self.

## A DAY IN THE LIFE

The understanding of the role of public symbolic interaction in the constitution of the self that both Mead and Vygotsky discuss at length gives us resources to explain what many consider the most important type of real, representation-hungry cognition: inner speech. Most of us spend large parts of the day speaking silently to ourselves. This is how we plan what to do, remember what we have planned, solve problems, and the like. This is what leads thinkers like Noam Chomsky to take language as having evolved primarily for thinking and only secondarily for communication. Our inner monologue is another undeniable case of real thinking. Unlike the sorts of imaginative and creative activities involved in thinging, our inner speech is hidden and inner. This might make it seem difficult to explain for the sort of nonrepresentational, embodied, and social understanding of the mind I have developed. In what follows, I will draw on Vygotsky, Mead, and some current thinking in the philosophy of science to account for inner speech.

Imagine a typical morning: You are at the breakfast table, drinking coffee. You look at the clock and realize you are late for work. You stand up and pat your pocket to see if you have your keys. You don't, so you wonder where they are. You usually leave them on your nightstand, so you decide to look there first. You walk toward the bedroom, and on the way there, you see the laundry pile and think that you need to do laundry tonight, which reminds you that you don't like the smell of the new laundry detergent you bought recently, which makes you wonder what detergent Satoshi uses that makes Satoshi's clothes always smell so good. You get to the bedroom, see your keys on the nightstand and grab them. On your way to the front door, you notice the dirty dishes. You consider loading the dishwasher, but you decide you can't right now, so you grab your bag to get ready to leave. As you grab your bag, you remember you are going to need a certain book today, so you open the bag to see if the book is in it. Inside the bag, you see your laptop and you notice, on the laptop case, the mashed core of an apple you ate the day before. While you bemoan being such a slob, you grab the apple core and throw it away in the kitchen. You wet some paper towels, bring them to your bag, and wipe your laptop case. You figure it looks clean enough, close your bag and rush out the door having forgotten to get the book. As you walk out the door, you see the emptied trash cans on the curb and you think about getting them, but you immediately rule that out: you can't deal with that right now or you will be late.

The story above illustrates the way our inner speech typically comes up in daily contexts: during our various activities, we think, reflect, understand, realize, decide, plan, and remember. In the midst of what we do and see, we think about what we do and see as well as about what is not present then and there—what we might see and do later or what we have done in the past. You see the clock and realize you are late, and you see the pile of dirty clothes and make plans to do laundry tonight: in both cases, there is something in your surroundings that guides your thinking. But thinking about something present can also lead us to thinking about things that are *not* present: this happens when you look for an object (e.g., trying to find your keys in the process of preparing to leave) and also when you reflect on your situation and imagine counterfactual ones (e.g., thinking about the laundry detergent you have right now and then thinking about other options and wondering what detergent someone else uses). Table 7.1 gives a breakdown of the story above emphasizing this intimate connection between thoughts

**TABLE 7.1** Relationship of occurrent thoughts to perception-action events

| | |
|---|---|
| Drinking coffee. See clock. | "8:15! Late for work!" |
| Stand up. Pat pants pocket. | *"Keys? No keys!"* <br> *"Where are they?"* <br> *"Usually I leave them on my nightstand"* |
| Walk toward bedroom. <br> Pass laundry pile. | "I need to do laundry tonight" <br> *"I don't like the way the new detergent smells. Satoshi always smells so good. I wonder what kind he uses."* |
| Continue into bedroom, see keys on nightstand. Grab keys. | "There!" |
| Walk toward door. See dirty dishes. <br> Grab bag. | "Nope. Too late." <br> *"I need that Edith Stein book. Where is it?"* |
| Look in bag. See laptop. Smell, then see mashed apple core. | "Ewww!" <br> *"Why am I such a slob? I ruin everything."* |
| Grab apple core. Go to kitchen. <br> Fling apple in sink. Wet paper towel. Wipe laptop case. | "Good enough, I guess." <br> "I'm late!" |
| Go out door. See emptied trash cans. | "I can't deal with that now." |
| . . . | . . . |

*Note:* Thoughts in italics are not directed at objects that are perceptually present (e.g., wondering where your keys are).

and specific perception-action events, also highlighting how thoughts relate to objects that are perceptually available or not. To be clear, in listing perception-action and thoughts separately, I do not mean to suggest that the two are independent and distinct in kind: on the contrary, thoughts are always associated with our active embodied engagement with the material and social world. Still, listing thoughts separately highlights the fact that even thoughts about objects that are not immediately available for perceptual inspection (e.g., the missing keys, on the nightstand in another room) can be useful for coping with aspects of the situation that *are* perceptually present (e.g., looking at the clock and realizing you are late).

## SCIENTIFIC MODELS AS ARTIFACTS

The account I want to offer of inner speech is inspired by a view that has become popular in the recent philosophy of science literature on models and simulations, according to which the various types of models that scientists use—including concrete scale models, mathematical equations and computer simulations—are best understood as tools, instruments, or artifacts. Traditional philosophical analysis has focused on explaining how models represent target real-world phenomena.[14] But some artifactualists question this traditional representationalist focus, and they propose that the task of elucidating how models represent target phenomena is secondary, and perhaps even unnecessary, compared to the task of understanding how models are used.[15] This artifactualist shift corresponds to a shift away from analyzing scientific models as "models of" the world toward understanding models as "models for" some purpose.

Particular scientific models are for some action in that they are designed to be manipulated in specific ways and to accomplish specific tasks. Besides hypothesis testing, explaining past events, and predicting future events, models might be used more generally for a variety of purposes, such as generating understanding, enabling interventions, designing new experiments, simplifying complex problems, and unifying apparently disparate phenomena. What is important is to understand what task a model is for, just as it is important if you want to understand what task any tool is for. Models are also for someone insofar as they are designed to be used by scientists with specific skills and interests, and to contribute to inquiry in specific research contexts. The Khepera robot has been useful for some modelers studying cooperative behavior in ants and phonotaxis in crickets,[16] yet it is of little use for the entomologist interested in the molecular genetics of ants or crickets. The "who for" and the "what for" of models are inseparable. Thus, scientific models mirror the case of ordinary tools: if we say that knives are for cutting, for example, it is implied that they are for animals capable of prehensile grip. Emphasizing the user in this way highlights how models, as tools, are always evaluated in terms of both their usefulness and their usability. Models have to be functional and effective toward their "what for;" that is, they have to produce the intended results.[17] But models also have to be workable and user-friendly: whether a model gives intended results or not depends not only on the intended goal but also on the intended user and on the user's ability to put the model to work. Thus, understanding models as "models for" naturally motivates thinking about models as tools that are used by someone to do something in some context.

For artifactualists, the idea that models are evaluated in terms of their usefulness and usability does not mean that they do not also represent phenomena in the world. There are many different accounts of the nature of the representation relation, and depending on which definition of representation you choose, the same model may or may not be properly described as representing some target. Artifactualism does not claim that scientific models never meet formal definitions of representation. Instead, the claim is that meeting formal definitions of representation is incidental to the purpose of models. It does not matter ultimately if a model "represents" some target (e.g., if a robotic cricket represents a real cricket): what matters instead is the role that the model plays in the activities, interventions, and understanding of scientists. For artifactualists, models are always "models for": they are tools that scientists design, construct, and manipulate to accomplish various goals; as such, models are more or less valuable depending on how engaging with the model and understanding it leads to success in a range of activities in a given scientific research context.

## INNER SPEECH AS TOOL USE

For Mead and especially Vygotsky, our abilities to engage in inner speech, to consciously think, come from internalization of cultural artifacts. Because of this, we can apply artifactualism from the philosophy of science to make sense of inner speech. Artifactualism about inner speech is a straightforward application of artifactualism about scientific models to the domain of the mind. Things we say to ourselves are tools, instruments, or artifacts that we create and use. Mirroring the artifactualist view of models as "models for" rather than "models of," thoughts are best understood as being "thoughts for" rather than "thoughts of." The purpose of inner speech is not to represent the world accurately: thinking is rarely if ever a matter of aimlessly taking stock of the elements present in our surroundings and generating or updating an internal inventory of the world. We don't note the position of every crease of every item in the laundry basket; we remind ourselves to do the laundry. The things we say to ourselves are tools for action, tools that are useful for accomplishing some goal.

The morning routine narrated above illustrates how inner speech can be useful for coping with perception-action demands. The acts of telling yourself that you are late and that you need to find your keys are useful tools for organizing your current actions and for adapting to your present circumstances (e.g., understanding that you need to look for your keys before you can leave the house). Making plans such as deciding to take care of the laundry, dishes, and trash later in the day is making tools for organizing future actions as well as current ones: having these episodes of inner speech can guide what you do at other points throughout the day (e.g., deciding to skip happy hour to go home and deal with the mess, or deciding to go to happy hour to avoid the mess) and can also inform what you do right now insofar as they enable you to draw your attention back to what is currently most pressing, namely, getting ready and leaving for work. Besides coping with our current situation and planning future actions, things we say to ourselves can be tools that we create and use for a variety of different purposes, including simplifying complex problems, predicting future experiences, explaining past experiences, categorizing experiences, confirming or disconfirming expectations, enabling adaptivity, generating understanding, and more.

Inner speech is also *for* some user; talking to yourself is a way to create tools that you can use. This point might seem trivial: unlike models, which are public objects that can be used by different people, inner speech is private. As tools, things we say to ourselves are not simply more or less useful for some end; they are also more or less usable for someone. Inner speech is not useful in some abstract and impersonal way: you need to know how you can use the things you say to yourself. Inner speech tends to satisfy this condition, and generally speaking, this is because we are successful perceiver-actors: contrary to what parts of the psychological literature might seem to suggest, perceptual illusions and motor failures are the exception, and we typically get by quite well. But we do sometimes have episodes of inner speech that we fail to use appropriately, and even some we do not know how to use well. For an example of failure to use inner speech appropriately, notice that in the narrative provided in this chapter, the inner sentence about the needed book did not lead to an appropriate action concerning the book.

Artifactualism about inner speech proposes that the purpose of saying things to yourself is not to represent the world accurately. That is, according to artifactualism, standard semantic notions like reference or content are not essential features of explicit, occurrent thoughts. Episodes of inner speech are ultimately tools for action, tools that we create, use, and evaluate based on how useful and usable they are. This does not mean that inner speech never meets formal definitions of representation. Instead, our claim is that meeting formal definitions of representation is incidental to the purpose of inner speech. It does not matter if our inner speech represents some target (e.g., if your plan to do laundry later accurately and fully corresponds to reality): what matters instead is the role that inner speech plays in your activities, interventions, and understanding. In talking to ourselves, we create and use tools to accomplish various goals, and thus instances of inner speech are more or less valuable depending on the extent to which they contribute to our successful coping with the world.

## SUMMARY

The purpose of this chapter has been twofold. First, it placed the views described in this book into deeper historical context. Along with phenomenological philosophers, the intertwined self is also in an intellectual lineage with thinkers such as James, Dewey, Mead, and Vygotsky. Second, it showed that the intertwined self does occasionally make aspects of our minds hidden and inner. Inner speech is not audible to others, so it is hidden. We can have secrets that no one else knows. The key insight of the pragmatists and Vygotsky is that these hidden, inner episodes depend on publicly used cultural tools—conversations, speech acts, gestures, and the like. We do have hidden, inner lives, but they are secondary. Our minimal selves and our minds are embodied and social. From artifactualism, we learn that we can understand these tools in terms of who is using them and what they are using them for. Our inner speech is tool use, and we use those tools to shape our ongoing actions; that our inner speech sometimes represents aspects of the world is not essential to our ability to use it to cope with the environment.

Artifactualism allows us to see that, from the point of view of the intertwined self, inner speech is a way of making tools to cope with situations. Those tools might sometimes represent the world accurately, but that is not what they are for and not what makes them tools.

*Chapter Eight*

# REORIENTING ETHICS AND POLITICAL THEORY AROUND THE INTERTWINED SELF

## INTRODUCTION AND RECAP

For a certain group of readers, my concern about treating cognition as dynamic and not as representational will remind them of the French philosopher Gilles Deleuze. For another, likely partially overlapping group, my insistence on the self as necessarily social will remind them of feminist political theory. At the same time, most cognitive scientists and philosophers of cognitive science will be unfamiliar with both. The purpose of this chapter is to explore the connections between intertwined self and certain figures in so-called continental philosophy and with some claims made by feminist philosophers and political theorists. To begin, I will draw a contrast between the intertwined self and the self as understood in liberal political theory, beginning with John Locke. Then I will discuss some rarely drawn connections between embodied cognitive science and continental philosophy. These will all be comparatively brief. I will devote most of the chapter to discussing ideas from feminist political theory, whose proponents developed a precursor to the intertwined self as part of their critique of patriarchal systems decades before any scientific evidence of the kind I have set out was available. I will argue that the phenomenological and scientific considerations are further evidence in favor of their case and the solutions they propose. This chapter comes with a warning: I am a novice at continental philosophy and feminist political theory. I will be relying, even more than before, on the writings of genuine experts.

Recall from chapter 1 that although Locke disagreed vehemently with René Descartes concerning epistemological issues, these disagreements stem from a shared commitment to understanding the mind as hidden, inner, and deeply disconnected from the surrounding world and other people—we could be in Newton Haven, surrounded by blanks. For Descartes, the chasm between the mind and world can be bridged because the mind is rich with innate knowledge, and this innate knowledge is the foundation of the mind's true internal representations of the external world. Locke denies both that the hidden, inner mind has innate structure and that it can form true representations of the external world. But each accepts what I have been calling the modern theory of the mind. Recall too that Locke brings this understanding of the mind to his highly consequential political work. What he calls a self in his *Essay Concerning Human Understanding* is the same thing as a mind; what he calls a person in his *Second Treatise on Government* is the same thing as the self. Locke then goes on to argue that a person (self, mind) is free to behave as they wish, to own property, and must consent to being governed. This understanding of a person is the one that becomes the center of Western politics; it is repeated in the US Declaration of Independence and was an inspiration for both the American and French revolutions of the late eighteenth century.

But given what I have been arguing in this book, this modern view of what it is to be a person is at best optional. The intertwined self is a philosophically sound and empirically supported understanding of the nature of the self, mind, and person. It is at odds with the view that Locke and Descartes develop. Given the importance of the modern view of the mind in Western political theory, the intertwined self is also at odds with much of Western political theory. In the rest of the chapter, I point to the political implications of the intertwined self, primarily by discussing some critical and feminist theorists who have developed similar views. I also discuss some of the responses they have made to the implied rejection of the foundations of Western political theory that it entails. My intention is that the scientific evidence I have presented in favor of the intertwined self should be seen as empirical confirmation of the work of these critical and feminist thinkers and in favor of the solutions they recommend.

To be clear, this chapter should be seen as especially tentative, a series of pointers to work that is rarely considered by cognitive scientists. As Francisco Varela put it in his initial foray into the ethical and political sphere, "What I have to say here must be taken in the spirit of adventure more than anything else. Yet it is terrain I am eager to explore."[1] My hope is that my tentative explorations will spur others to further study of the relations between the embodied, social self and moral and political theory.

## ASSEMBLAGES AND CYBORGS

As noted above, some readers will have seen the focus in this book on dynamical systems and the refusal to understand cognition in terms of representation and wondered why there has not yet been a discussion of Deleuze. Other readers will wonder why I am bothering to discuss Deleuze at all, given his notoriously difficult writing style and absence from the debates in the cognitive sciences. In this section, I hope to satisfy both types of readers by putting the ideas presented so far in the context of Deleuze's thought as clearly and carefully as I can, focusing on ideas from three of his major works: *Difference and Repetition*, *Expressionism in Philosophy: Spinoza*, and *A Thousand Plateaus* (coauthored with psychoanalyst Félix Guattari).[2] I also rely heavily on secondary literature, especially work by Moira Gatens, Daniel Smith, John Protevi, and Adrian Parr.[3]

For Deleuze, the history of philosophy is a series of "images of thought," a series of views about the relationship between knowing subjects and known objects. For Plato, for example, knowledge was a relationship between a fixed eternal soul and a fixed eternal form. For Descartes and the modern theory of mind, the relationship is between a hidden, inner mind that represents objects in the public, external world; the subject has knowledge when it represents the object correctly. Deleuze's overall philosophical project is to replace images of thought that take belief and knowledge to be any type of representation relationship between a fixed subject and a fixed object and to replace it with a flat ontology of becoming. As we will see, this is a rejection of the modern theory of the mind and one that is fully in line with the intertwined self.

As in Deleuze, the phenomenological, ecological, and enactive views that form the background for the intertwined self are antirepresentational; like Deleuze's views, they are theories of thought without an image. On representational theories, like the modern theory of the mind, knowledge is a matter of identity between what is represented in the mind and what is in the world. This is what it is for representations to be accurate. Deleuze argues that this image of thought makes it impossible to understand difference other than negatively, as a failure of identity. Identities can only hold between fixed entities. To claim that my thought is true, we would have to say that the world as I represent it right now is identical to the world right now. This would require that both my representation and the world are frozen, at least for the moment that my representation is true. Deleuze rejects this, arguing that everything is in constant, dynamical flux. In fact, prominent among the many forms of the word "difference" that appear in *Difference and Repetition* is "differentiation," by which Deleuze means the differential function $\dot{x}$ of dynamic systems theory. That is, like proponents of the ecological psychologists and enactive approaches, not to mention the anthropologists Tim Ingold and Lambros Malafouris discussed in chapter 6, Deleuze adopts an ontology according to which the world exists in constant flux. Deleuze's focus on differentiation also makes the methods of radical embodied cognitive science, discussed in chapters 4 and 5, an appropriate approach in scientific psychology. In this view, nothing is fixed enough to be a representation or represented; everything is always in the process of becoming something else.

This constant state of flux is Deleuze's ontology of becoming. It is *flat* ontology of becoming because it makes no distinctions between subjects and objects, mind and world. Instead of subject and object, Deleuze talks about *assemblages*, dynamic and contingent groupings of heterogeneous elements that come together temporarily to produce a particular form or entity; Deleuze and Guattari sometimes call assemblages machines, especially when they are focusing on them as active and productive. (I will use the terms interchangeably.) The elements of assemblages are not limited to material objects but also include relations, flows, and processes. Assemblages are not limited to humanmade artifacts; not all Deleuzian machines are machines in the usual sense. Bicycles are machines, but so are organs and organisms, and so are organisms-riding-bicycles, and so are languages, conceptual systems, and social institutions. It is easy to read the discussion of synergies, including interpersonal synergies, as a scientific approach to Deleuzian machines. For Deleuze, no part of a machine, even when the machine includes an organism or its brain, is privileged as a locus of experience or supplier of subjectivity. As we saw with Ingold and Malafouris in chapter 6, Deleuze does not locate agency just in organisms. All the elements of assemblages have power to affect and be affected by the other elements.

The traditional role of the subject or self or mind, as in the modern theory of the mind, is played by a disjunctive temporal collection of what Deleuze calls larval subjects. *Larval subjects* are assemblages, and like all assemblages, they are always in the process of becoming something new. Also, as assemblages, larval subjects are collections of biological and nonbiological elements, with no elements privileged as the locus of experience. This leads Protevi to align Deleuze's views with the enactive and complex systems approaches discussed above and to categorize Deleuze's view as a kind of biological panpsychism, in which everything that forms an assemblage with a biological organism is equally a site of experience.[4] For current purposes, larval subjects are often machines that include multiple organisms as elements. The human-tool and human-human synergies discussed in chapter 5 are all larval subjects. Because a subject or self is a collection of larval subjects, the machines that compose the self will have tools and other humans as elements. These larval subjects are the elements for any self, even a minimal or ecological self. For Deleuze, even a minimal self is embodied and social.

Donna Haraway, in discussing *A Thousand Plateaus*, says, "I am not sure I can find in philosophy a clearer display of misogyny."[5] Although there are crucial differences in their overall views, what Haraway calls cyborgs are also varieties of assemblages in Deleuze's sense.[6] For Haraway, a *cyborg* is an entity composed of interdependent biological, cultural, and technological parts. In a later work, Haraway describes cyborgs as follows:

> Cyborgs are not machines in just any sense, nor are they machine-organism hybrids. In fact, they are not hybrids at all. They are, rather, imploded entities, dense material semiotic

"things"—articulated string figures of ontologically heterogeneous, historically situated, materially rich, virally prolifer-
ating relatings of particular sorts, not all the time everywhere,
but here, there, and in between, with consequences ... Cyborgs
are constitutively full of multiscalar, multitemporal, multima-
terial critters of both living and nonliving persuasions.[7]

Haraway's intent with cyborgs is to blur the distinctions
between body and world, human and nonhuman, animal and
machine, and nature and culture.

Haraway explicitly contrasts her view with the enactive idea
of autopoiesis, discussed in chapter 2. Cyborgs are not just living things; they are "constitutively full" of living things. As such,
they do not just engage in autopoiesis, or self-making. Instead,
they engage in *sympoiesis*, or making with.[8]

> *Sympoiesis* is a simple word; it means "making-with."
> Nothing makes itself; nothing is really autopoietic or self-
> organizing. In the words of the Inupiat computer "world
> game," earthlings are *never alone*. That is the radical impli-
> cation of sympoiesis. *Sympoiesis* is a word proper to complex,
> dynamic, responsive, situated, historical systems. It is a word
> for worlding-with, in company. Sympoiesis enfolds autopoi-
> esis and generatively unfurls and extends it.[9]

Haraway's focus on sympoiesis is not in conflict with modern
enactive theory, which has also recognized that autopoiesis is
not sufficient and replaced it with autonomy: both biological individuals and episodes of participatory sensemaking are

autonomous. Haraway's point in introducing sympoeisis is to deny that the biological individual is the appropriate unit to understand the nature of life and living things. A central feature of Haraway's views is the discussion of cyborgs that comprise humans and companion animals. The intertwined self is also intertwined with nonhumans.

Like Deleuze, Haraway's aim is disruptive: with cyborgs, we can leave behind traditional dualisms and essentialisms. Unlike Deleuze, Haraway's aim is explicitly feminist. Cyborgs are complex and fluid, which puts them in conflict with essentialism about gender and opens up possibilities for feminist social and political theorizing. The intertwined self is a cyborg. This suggests that the intertwined self has feminist consequences. We explore these in the next section.

## FEMINISM, ONTOLOGY, AND VALUE

Gatens points out that Western philosophy has historically excluded and erased women explicitly.[10] Although she does not use the term, what she identifies as the cause of this exclusion and erasure is the modern theory of the mind. She traces the conception of a person in political philosophy from Locke to Jean-Jacques Rousseau, to John Rawls, pointing out that each of these thinkers (and many of those responding to them) understands being a person as being an idealized version of a man. A person, according to these thinkers, is autonomous, rational, and self-interested—essentially calculating machines, working to maximize their own well-being.

The male subject is constructed as self-contained and as an owner of his person and his capacities, one who relates to other men as free competitors with whom he shares certain politico-economic rights. While he has rights to privacy and self-improvement, he relates to women as though they were a natural resource and compliment to himself. The female subject is prone to disorder and passion, as economically and politically dependent on men, and these constructions are justified by women's nature.[11]

The supposedly sexually neutral person is this male subject. Gatens quotes Rousseau's *Émile*: "But for her sex, a woman is a man." For Rousseau, a man is a person in the sense just described; a woman's body interferes with her abilities to be a person in this sense. Women's bodies, according to Rousseau, make them emotional rather than rational, confined to the realm of nature and excluded from culture, relegated to the private home and excluded from participation in public life. All humans have a hidden, inner mind, but a woman's body interferes with its expression in action.

The impact of the modern theory of mind, as hidden, inner locus of rationality, which we now can see as an idealized version of a man's mind, is especially evident in Rawls's influential *A Theory of Justice*.[12] Rawls asks readers to imagine a communion of minds, in what he calls "the original position": not in a body and meeting behind a "veil of ignorance," behind which they will not know which body they will return to, including crucially the behavioral, cultural, and economic position of that body. Behind this veil, the disembodied minds use their perfectly rational self-interest to debate and agree on a set of rules and regulations according to which they will be governed when they return to their bodies. No rational, self-interested mind would argue for any kind of unfair division of resources or opportunities because it would not know whether it would be going back to the body of a member of a class it has just argued to disadvantage. Rawls concludes that disembodied rational minds would agree that all assets and opportunities must be divided equally; this is the only way to serve their own interests. For decades, political theorists have found this thought experiment interesting, even inspiring—the book has been cited more than 100,000 times according to Google Scholar. The thought experiment relies unfortunately on the very conception of mind that Gatens and I have argued against. From the point of view of feminist political theory and the intertwined self, the idea of a rational, disembodied mind is nonsense.

Lorraine Code offers a replacement understanding of the mind in the form of a relational ontology of persons, designed to challenge the centrality of autonomy and to value prototypically feminine ways of being, both morally and epistemically.

The autonomous moral agent is the undoubted hero of philosophical moral and political discourse: the person—indeed, more accurately, the *man*—whose conduct and attributes demonstrate the achievement of moral maturity and goodness . . . Positing achieved autonomy as the mark of moral maturity has had the effect of withholding approval from people whose conduct, in fact, is often praiseworthy. Women, children, Blacks, the aged and the ill are just some of many examples: people whose politically constructed and enforced lack of autonomy excludes them from full moral agency, rendering them dependent and subject to paternalistic control.[13]

Code argues that this conception of autonomous moral agents has consequences beyond excluding minorities, women, and children from moral agency. First, it makes autonomy closely aligned with an individualism that views interdependences, like those between romantic partners or between caregivers and children, as diminishing autonomy and therefore personhood. Second, it emphasizes the otherness of all other people, even those with similar social, cultural, and economic status, casting them as essentially opaque and alien. That is, it puts us back in Newton Haven, where those around us might be blanks and, even if they are not blanks, they are fundamentally competitors.

Drawing on Annette Baier's conception of "second persons,"[14] Code replaces autonomous man with "a conception of cognitive agency for which intersubjectivity is primary and 'human nature' is ineluctably social."[15] From this point of view, developing the skills and traits that make one a person requires having spent a significant amount of time dependent on other persons. Being a person requires having been, and indeed continuing to be, a second person. "A human being could not become a person, in any of the diverse senses of the term, were she or he not in 'second person' contact from earliest infancy."[16] On this view, persons are creations of other persons, and continued interdependence is in no way inimical to agency, even for adults. "If persons are essentially second persons, there can be no sense in assuming that they grow naturally to autonomous self-sufficiency, only then—perhaps cautiously, incidentally, or as an afterthought—to participate in intimate relationships."[17] This view is fully in line with what I have called the intertwined self.

This view of what it is to be a person has ethical consequences, which Code and others develop. I will point to these in the next few sections of this chapter. In the meantime, I want to point out something that many readers will have noticed already: the works I have discussed so far in this chapter are not recent. Deleuze came to a view of a person quite similar to the intertwined self via a critique of twentieth-century capitalism. Feminist thinkers like Haraway, Gatens, and Code got to the same place as part of a critique of the patriarchy. These are explicitly political critiques. Until recently, philosophers in the ecological-enactive tradition have shied away from explicitly bringing political considerations into their theorizing. Only in the second decade of the twenty-first century was the connection to feminist and critical literature made. Among those who have made explicit connections between the ecological-enactive cognitive science and feminist thinking are philosophers Mason Cash, Elena Cuffari, Michelle Maiese, Hanne de Jaegher, Sara Heinämaa, Nick Brancazio, Laura Candiotti, Anya Daly, Janna van Grunesven, Juan Loaiza, Sam Liao, Bryce Huebner, and Vanessa Carbonnell, along with psychologists Julia Blau, Alexandra Paxton, and Mikayla Weston.[18] We will meet several of these thinkers again in the remainder of this chapter.

## RELATIONAL AUTONOMY

Feminist critiques of autonomous man do not mean that we can do away with the concept of autonomy. Catriona Mackenzie and Natalie Stoljar note that, from a feminist point of view, "the concept of autonomy is inherently masculinist," but also that, although "feminist critiques of autonomy have identified serious theoretical and political problems with some historical and contemporary conceptions of autonomy, the notion of autonomy is vital to feminist attempts to understand oppression, subjection, and agency."[19] Thus, in light of critiques like those from Gatens and Code, feminist thinkers have elaborated what they call relational autonomy. *Relational autonomy* is an attempt to acknowledge simultaneously both that becoming autonomous requires being in relationships with other people and that the autonomy of individuals can be both enhanced and limited by their race, gender, or social position. That is, being autonomous is never absolute and always depends on other people and the cultural, material, and technological environment. Relational autonomy is the autonomy of second persons. It is also the kind of autonomy available to intertwined selves.[20]

Mason Cash draws the connection between socially distributed cognition and relational autonomy, developing a view of cognitive agency in which the autonomy of agents is not a property of biological individuals alone.[21] "Many—principally feminist—scholars have critiqued accounts of self, agency, and autonomy that view the self as a Cartesian mind or an unencumbered individual who thinks and acts independently from any external influence."[22] Just as individual human cognition takes place in the context of and depends on tools and other entities in the environment, including other people, individual human autonomy always occurs in relation to and depends on the environment and other people.

> An autonomously "intelligent" agent is one who critically engages the cognitive tools around them, one who selects, endorses, and uses effective cognitive tools, who replaces, refines, or augments less effective cognitive tools, and who selectively incorporates these social, relational, technological, environmental, and bodily resources into their sense of who they are, what they know, what they want, and what they can do.[23]

If cognitive systems are larger than biological bodies, so too are agents. The autonomy that agents possess makes sense only in the context of their technical and cultural environments, including other people. This relational form of autonomy also inherits the biases and oppression extant in the culture.

In her book *Autonomy, Enaction, and Mental Disorder*, Michelle Maiese develops a theory of relational autonomy specifically in the context of the ecological-enactivist approach to cognitive science.[24] Maiese argues that our most basic experiences of the world are affective, that is, simultaneously about the world and valenced. She calls this *affective framing*; when we have these experiences of other living things, they are what I called sensorimotor empathy in chapter 3. To be clear, Maiese would reject what I have been calling the intertwined self. In the language used in debates over relational autonomy, she thinks that autonomy is relational in that agency is causally embedded in the material and social world but that it is not constituted by relations to the material and social world. Maiese argues that taking agency to be constitutively social is in tension with some of the core ideas of the enactive approach. I have argued, following Hanne de Jaegher, Ezequiel Di Paolo, Miriam Kyselo, Anna Ciaunica, and Sanneke de Haan, that from an ecological-enactivist point of view, even the minimal self is partly constituted by interactions with the world and with others.

Despite this disagreement, we can adopt Maiese's understanding of autonomy as relational, largely because she couches it in terms of dynamical systems theory. As we saw in chapter 5, self-organizing dynamical systems can span body and environment and can also include components of other humans; that is, they support the intertwined self. An agent exhibits relational autonomy when they can, sometimes in spite of their social position, adopt new adaptive ways of engaging the world. According to both Maiese and the intertwined self, organisms are collections of self-organizing dynamical systems that are self-maintaining but subject to constraints and perturbations from their environment; according to the intertwined self, but not Maiese, these dynamical systems extend beyond biological individuals. Either way, habits of action develop in an environment and tend to maintain themselves in the face of certain perturbations of their activity. Thus, for example, autonomy-restricting behavioral tendencies will tend to maintain themselves, all things considered. However, agents are *collections* of such self-organizing habits of action, which means that those habits tend to enable and constrain one another. This interconnection of habits is the key to the dynamical approach to relational autonomy: an agent might have habits of action that inhibit their autonomy that develop in virtue of their marginalized social position, but they also have other habits, related, say, to critical thought or resilience, that can disrupt the activity of the autonomy-inhibiting habits that constitute the same agent. That is, because they are components of one another's environments, the self-maintaining tendency to speak up interacts with and might overcome the self-maintaining tendency to accept blame for the actions of others.

## VIRTUE, CARE, EMPATHY

If the intertwined self suggests that the rational, autonomous moral agent is a gendered philosophical fiction, it also suggests that we need to reject traditional moral theories, like consequentialism and deontology, which define morality in terms of rationality and universal principles. In consequentialism, morality is a matter of rationally calculating net costs and benefits of actions and always acting to maximize overall benefits and minimize overall costs; in deontology, it is a matter of rationally evaluating situations and applying appropriate universal rules of behavior. The problem with these views is that they are impartial. For a consequentialist, it doesn't matter who benefits as long as the overall benefits are maximized; for a deontologist, the rules are applied impartially, with the very same rules applied no matter who the individuals involved are. These traditional moral theories force us to treat other humans as interchangeable. It would be bizarre for a mother to treat her child using the same principles she would apply to strangers, but this is what is required by traditional moral theories. Intertwined selves are second persons, necessarily enmeshed in close personal relationships; they require a moral theory where others are treated as particular individuals for some of whom we feel friendship, loyalty, and love. We are lucky that there are alternative theories that are not wedded to impartiality, two of which have already been of interest to both feminist and ecological-enactive theorists. The first of these, virtue ethics, can be traced back to Aristotle and Confucius; the second, the ethics of care, is more modern, initially developed by Carol Gilligan in the late twentieth century.

The two key concepts in virtue ethics are virtue and practical knowledge.[25] Virtue here means exactly what it does in normal speech: a virtue is a praiseworthy character trait and a tendency to act in a way that expresses that trait; having practical knowledge is knowing which virtues to express and when to express them. For example, having the virtue honesty is to tend to tell the truth, all things considered, but also having the practical knowledge to know to tell the truth nearly all the time, but possibly not when someone asks, "Isn't my baby adorable?" Virtue theory focuses on the person rather than on individual actions. A morally good person cultivates and appropriately acts on virtues such as honesty, wisdom, compassion, courage, and so on. In this view, being a good person is a matter of moral education and character development, not the application of rules or the calculation of costs and benefits. Eranda Jayawickreme and I have drawn explicit connections between the ecological approach in psychology and virtue ethics.[26] Recall that an affordance is a relationship between a creature's abilities and the situation in the environment. A person with the ability to walk and a street with no moving cars together afford street crossing. Virtues can be understood as a subset of our abilities, those which enable us to act in moral situations. The relations between this subset of abilities and situation in which they might be applied are moral affordances.

The practical wisdom part of virtue ethics is parallel to the distinction between affordances and invitations, which was discussed in chapter 2. At any moment, many thousands of actions are afforded to a human agent, but they are aware of almost none of them—almost none of the affordances invite their action. Erik Rietveld and colleagues distinguish between the field of relevant affordances that invites action and the much larger landscape of all the things that are afforded to a person. Being able to make this distinction between affordances and invitations in a moral context is practical wisdom. I have the ability to help someone who is lying on the ground to get back to their feet. Exercising this ability at the grocery store with someone who has fallen is acting on a moral affordance and an expression of the virtue kindness; exercising it with people lying on blankets at the beach would be unwelcome at best. To have practical wisdom is to know the difference between moral affordances, situations in which you can help, and moral invitations, situations in which it is appropriate for you to help. Having practical wisdom in this sense does not require impartiality. A soiled diaper invites my action when it is my child's diaper, and although someone else's child's diaper might afford changing given my abilities, it does not invite it except in special circumstances, such as when I am the babysitter.

This lack of impartiality is also a feature of the ethics of care. The ethics of care is part of the feminist philosophical tradition, originating in the work of the psychologist Gilligan.[27] In the 1980s, Gilligan noticed that participants who were women and girls routinely scored lower than participants who were men and boys on Lawrence Kohlberg's moral maturity scales. Rather than treat this as a systematic moral immaturity of women and girls, Gilligan pointed to gender differences in moral reasoning and biases in Kohlberg's scales. Men and boys, including Kohlberg, who designed the moral maturity scales, tend to reason about morals in terms of impartiality and rationality—they tend to be consequentialists or deontologists; in contrast, women and girls tended to reason about morals in terms of caring and relationships with particular others. This, Gilligan argued, is not a matter of women and girls being somehow deficient in moral reasoning but rather that they reason differently, *In a Different Voice*, as the title of Gilligan's book has it.[28] This has been the dominant view in feminist ethics circles for several decades now.[29] It is also perfectly in line with the intertwined self.

The core of the ethics of care is that humans are second persons and caring for others is an essential aspect of what it is to be human and to live a moral life. Rather than focusing on the well-being of abstract others, the ethics of care emphasizes the moral responsibility to respond to the needs and vulnerabilities of family, friends, and communities. The ethics of care, therefore, rejects the notion of impartiality in moral reasoning—it is wholly appropriate to value the well-being of those with whom we are in close personal relationships over the well-being of abstract others. It encourages individuals to consider the particular needs and circumstances of others and to engage in caring practices that promote well-being, growth, and mutual flourishing. Since Gilligan introduced the ethics of care, it has been argued that a focus on caring is not specific to women and girls but is central to the moral reasoning of marginalized groups more generally.[30] The ethics of care also critiques traditional ethical theories for their failure to acknowledge and address the unique experiences and perspectives of marginalized groups, including women and caregivers, among others. This is not merely conceptual: Anjali Dutt and Danielle Kohlfeldt argue that thinking of community relations in terms of care ethics has potential liberatory consequences for marginalized groups.[31]

The relationship between virtue ethics and the ethics of care is disputed, with some theorists arguing that the approaches are naturally allies and some arguing that they are incompatible. This is hardly the place to settle this debate, but it is worth pointing out that Mark Órnelas has developed an argument parallel to the one Jayawickreme and I offered for a connection between ecological psychology and the ethics of care. Caring for others, Órnelas argues, requires sensitivity to situations that afford and invite caring behavior.[32] The philosophers Janna van Grunesven and Juan Loaiza separately argue that the ethics of care is the appropriate ethics for the enactive approach.[33] Both van Grunesven and Loaiza focus on the centrality of participatory sensemaking in enactive theory. In participatory sensemaking, we form autonomous units with other individuals, in effect treating them as second persons, which invites a commitment to an ethics of care.

The point of the foregoing is that the intertwined self is incompatible with traditional consequentialist and deontological ethical theories. Both virtue ethics and the ethics of care are committed to the centrality of empathy. So too is the intertwined self. Empathy is the key to being with others, to being a second person. As we noted in chapter 3, empathy here is not primarily a matter of reasoning about the mental states of others to imagine what they might be thinking; instead, it is a matter of direct social perception, a kind of active bodily engagement with others that I have called sensorimotor empathy. Sensorimotor empathy is affective; that is, it is more basic than a distinction between cognition and emotion. From the point of view of traditional ethical theorists, this would make sensorimotor empathy inappropriate for rational, moral deliberation about which rules to apply or which act has the best consequences. Some thinkers, like Paul Bloom,[34] have argued that even cognitive empathy, which includes explicit theorizing about or simulations of the mental states of others, can have negative impacts for moral reasoning because moral reasoning should be rational and impartial. From the point of view of the intertwined self, that is, from the point of view of virtue ethics or the ethics of care, this is simply incorrect. We do need to take personal relationships into account, and we owe different things to our children, friends, and lovers than to those we have never met.

## IN FAVOR OF ACTIVISM

Most academic disciplines in the humanities and social sciences have an explicit activist attitude, where they view themselves as generating basic knowledge about the world and simultaneously working to empower marginalized groups. The physical and life sciences, including the cognitive sciences, tend not to share this activist bent—and surprisingly, neither does academic philosophy. Like physical and life scientists, most philosophers think of themselves as committed professionally to pursuit of the truth but not to social change. The alliances I have drawn in this chapter between the interactive self and feminist and continental philosophers suggests that this is a mistake. The specific moral and political ideals I focus on are not the only ones I might have chosen to discuss, of course. The point of the chapter is to start a conversation about the connections between philosophical and scientific theorizing and our moral and political ideals. Philosophers and cognitive scientists are in positions to do basic research aiming at the truth and also to promote positive social change. Failing to try to do both is an abdication of responsibility. As Karl Marx puts it in the *Theses on Feuerbach*, "Philosophers have hitherto only interpreted the world in various ways; the point is to *change* it."[35]

*Chapter Nine*

# CODA

## BLANKS AMONG US

**IN THE 2020s,** we are surrounded by blanks in the form of artificial intelligence (AI). For example, ChatGPT and other large language models (LLMs) generate clear and convincing prose on almost any topic; they can be useful for research; they generate computer code that compiles and runs; and ChatGPT has been judged by patients to be as medically accurate as and more empathetic than human physicians. Of course, ChatGPT is far from perfect. For example, despite valiant efforts by their creators, it is easy to reveal the racial and sexist biases that infect the training sets of LLMs, which are scraped from the internet.[1] ChatGPT is also prone to doing what its creators spin as hallucinating, that is, making things up whole cloth. LLMs could be likened to politicians running for office. Both have a ready answer to any question; both also tend to make things up. Politicians are intelligent and so are LLMs—even if both always require fact checking. Nonetheless, current LLMs and other forms of AI are blanks. They are not like us. They are not intertwined creatures.

I have argued that a basic human characteristic is that we are intertwined creatures: living, biological creatures who are partly constituted by the material, social, cultural, and technological environment. As living creatures, we have metabolic needs that are specific to the nature of our bodies. We have motor and perceptual systems tuned to help us to meet those metabolic needs. Frogs, for example, eat flies, and they are masters at detecting and catching them, just as modern humans are good at finding grocery stores and restaurants. The ways in which living, biological creatures like us perceive and move are tightly intertwined; perceiving the world is for guiding action and often involves action. To experience what is going on around us, humans and other animals move: we move our eyes, we crane our necks, and we walk over to get a better look at things. This activity is not ancillary to seeing what is around us; it is part of the seeing. In other words, from the embodied point of view I have developed in this book, human vision is not the responsibility of the eyes or a brain region. The thing that sees the world includes a brain and eyes, of course, but those eyes are always moving and are set in a head on a neck on the torso of an animal that is also moving. This tight connection between experiencing and acting is a central feature of human life and is something that current AI lacks entirely. Current AI programs are not embodied and, because of this, not like us.

AI is also not living, with metabolic needs like we have. As we noted in our discussion of the enactive approach in chapter 2, our metabolic needs imply a built-in positive or negative valence to our experience of situations in the world. We have needs and, because of this, some situations are experienced as being better than others. Even single-celled organisms respond differently to different concentrations of specific chemicals in their milieus, working to keep conditions within the constraints required for their viability. Human cognition is, at its root, a set of tools that we use to keep ourselves alive, and this is why we experience some situations as good and some as bad. We humans are motivated to be warm and fed and loved. This motivation pervades our experience and infects even the most apparently disinterested cognition.

I have also argued that the lives of humans are necessarily social. We begin our lives inside another human and are born utterly helpless. We develop in a world alongside other humans, especially with immediate caregivers who provide us with the warmth, food, and love required to keep us going. We can learn language more quickly and efficiently than AI can because, for us, words come along with facial expressions, intonations, gestures, and the context of temporally extended interpersonal interactions. We also live in specific cultural and technological settings that shape and enable our activities. As intertwined creatures, we navigate complex interpersonal situations, and manage our facial expressions, intonations, and gestures so that they fit the context and our role in it. We enter the lecture hall differently when we are the lecturer than we do when we are audience members; we enter the lecture hall differently in Tokyo than we do in Toronto.

Combined, these differences between AI and humans point to a way of living and being intelligent that is founded on what

John Haugeland called "giving a damn."[2] Intertwined creatures like us give a damn. Part of giving a damn is being committed to maintaining one's own existence and relationships with the world and others; another part is caring whether one is hallucinating or telling the truth. ChatGPT is incapable of caring about these things because it is not an embodied creature, alive and participating in the world that it generates text about. ChatGPT is a blank. To be sure, future AI models (maybe incorporating today's LLM technology) might be built so they do give a damn. What we learn from today's LLMs is that there is more than one way to be intelligent. Our way of being intelligent is not the only way and not their way. This takes nothing away from LLMs as a technological achievement but makes clear that they currently are not much help for answering scientific questions about the intelligence of humans and other animals. There is still plenty of work for cognitive scientists to do. Doing that work will require careful attention to our being intertwined—living, moving, social, and enculturated—creatures, creatures that give a damn.

# APPENDIXES

*I was on a long circuitous path, with not enough music and too much math.*

Winechuggers, "Long circuitous path"

# APPENDIX FOR PEOPLE WHO LIKE MATH

**HERE ARE BRIEF TECHNICAL NOTES** on some of the empirical results described in chapters 4 and 5.

## CHAPTER 4: FISH FIN COORDINATION

One could model Erich von Holst's observations of fish fin coordination with the following coupled equations:

$$\dot{\theta}_{fin1} = \omega_{fin1} + Magnet_{2 \to 1}\left(\theta_{fin2} - \theta_{fin1}\right)$$
$$\dot{\theta}_{fin2} = \omega_{fin2} + Magnet_{1 \to 2}\left(\theta_{fin1} - \theta_{fin2}\right),$$

where $\omega$ is the preferred frequency of each fin and $\theta$ is the current frequency of each fin. In English, these equations say that the change in frequency of oscillation of each fin is a function of that fin's preferred frequency and the magnet effect exerted on it by the other fin.

There are three things to note about these equations. First, the first and second terms in the equations represent the maintenance and magnet effects, respectively. Second, the equations are nonlinear and coupled. You cannot solve the first equation without solving the second one simultaneously, and vice versa. Third, with these equations, you can model the behavior of the two-fin system with just one parameter: the strength of the coupling between the fins, that is, the strength of the magnet effect.

### CHAPTER 4: FINGER WAGGING

Hermann Haken, Scott Kelso, and Herbert Bunz modeled finger-wagging behavior using the following equation, a potential function:

$$V(\phi) = -A\cos\phi - 2B\cos 2\phi,$$

where $\phi$ is the relative phase of the oscillating fingers and $V(\phi)$ is the energy required to maintain relative phase $\phi$. Relative phase is a collective variable, capturing the relationship between the motions of the two fingers. In-phase tapping, with both fingers making the same movements at the same time, has relative phase of 0, while out-of-phase tapping, with fingers making syncopated movements, is relative phase $\pm\pi$. The transition from relative phase $\pi$ to relative phase 0 at higher metronome frequencies is modeled by the derivative of relative phase:

$$\dot{\phi} = -A\sin\phi - 2B\sin 2\phi.$$

In these equations the ratio $B/A$ is a control parameter, a parameter for which small quantitative changes to its value lead to large qualitative changes in the overall system. These equations describe all the behavior of the system.

Don't worry if you don't understand the equations. The key parts of this and many other dynamical explanations is graphical. The behavior of the coordinated multilimb system can be seen in figure A1.1, which shows the system potential at various values of the control parameter $B/A$. At the top of the figure, when the metronome frequency is low and the control parameter value is near 1, there are two attractors: a deep one at $\phi = 0$ and a relatively shallow one at $\phi = \pm\pi$. As the value of the control parameter decreases (moving down the figure), the attractor at $\pm\pi$ gets shallower, until it disappears at .125. This is a critical point in the behavior system, a point at which qualitative behavior changes.

### CHAPTER 5: 1/F NOISE

It has long been known that self-organized critical systems exhibit a specific, fractal variety of fluctuation called $1/f$ noise, or pink noise.[1] $1/f$ noise or pink noise is a kind of not-quite-random, correlated noise, halfway between genuine randomness (white noise) and a drunkard's walk, in which each behavior is constrained by the prior one (brown noise). $1/f$ noise is often described as a fractal structure in a time series, in which the variability at a short timescale is correlated with variability at a longer timescale. Interaction-dominant dynamics predicts that this $1/f$ noise would be present. As discussed in

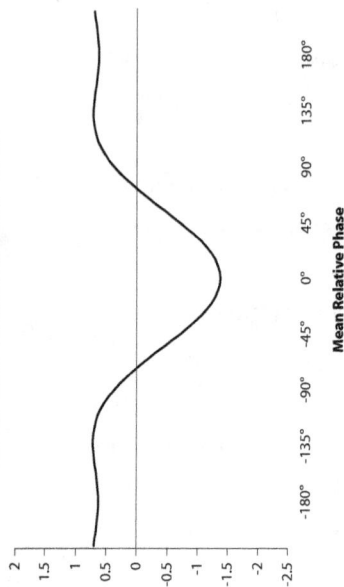

**Haken-Kelso-Bunz Model: b/a = 1**

Mean Relative Phase

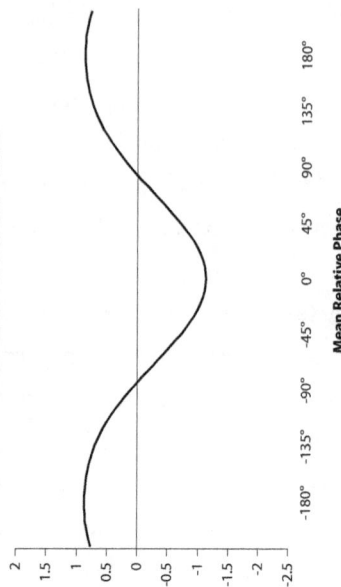

**Haken-Kelso-Bunz Model: b/a = 0.375**

Mean Relative Phase

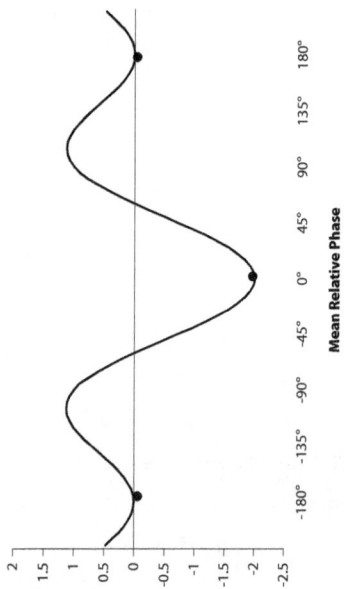

**Haken-Kelso-Bunz Model: b/a = 0.625**

Mean Relative Phase

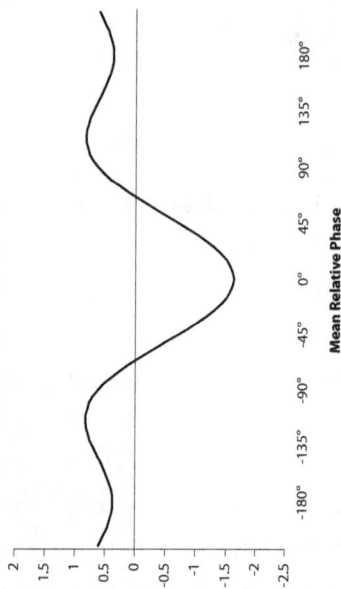

**Haken-Kelso-Bunz Model: b/a = 0.125**

Mean Relative Phase

**FIGURE A1.1** The potential function as modeled by Haken et al. *Top left:* one deep and one shallow attractor; *top right* and *bottom left:* the shallow attractor gets more shallow; *bottom right:* one deep attractor. Image created by author from Hermann Haken, Scott Kelso, and

chapter 4, the fluctuations in an interaction-dominant system percolate through the system over time, leading to the kind of correlated structure to variability that is $1/f$ noise. Here, then, is the way to infer that cognitive and neural systems are synergies: exhibiting $1/f$ noise is evidence that a system is interaction dominant; interaction-dominant systems are synergies. The mounting evidence that $1/f$ noise is ubiquitous in human physiological systems, behavior, and neural activity is also evidence that human physiological, cognitive, and neural systems are interaction dominant, which in turn is evidence that they are synergies.

### CHAPTER 5: HUMAN-TOOL SYNERGY

In detrended fluctuation analysis (DFA), which Dobromir Dotov and colleagues used to analyze hand-on-mouse movements, the one-minute time series is divided into temporal intervals of sizes varying from one second to the size of the whole time series. For each interval size, the time series is divided into intervals of that size, and within each of those intervals, a trend line is calculated. The fluctuation function $F(n)$ for each interval size $n$ is the root mean square of the variability from the trend line within each interval of that size. $F(n)$ is calculated for intervals of $1, 2, 4, 8, 16 \ldots$ second intervals. As you would expect, $F(n)$ increases as interval size increases. The values of $F(n)$ are plotted against the interval size in a log-log plot. The slope of the trend line in that log-log plot is the primary output of DFA, called a Hurst exponent. Hurst exponents of approximately 1 indicate the presence of $1/f$ noise, with lower values indicating white noise and higher values indicating brown noise. The raw data is folded, spindled, and mutilated to determine whether the system exhibits $1/f$ noise and can be assumed to be interaction dominant.

As predicted, the hand-mouse movements of the participants exhibited $1/f$ noise while the video game was working correctly. The $1/f$ noise decreased, almost to the point of exhibiting pure white noise, during the mouse perturbation.

Using DFA to claim that a system is interaction-dominant has been controversial. See the meta-appendix.

## CHAPTER 5: INTERPERSONAL SYNERGY

To show that the players in the video game formed an interpersonal synergy, Patrick Nalepka and colleagues relied on dimensional compression, another defining characteristic of a synergy.[2] Dimensional compression results from coupling degrees of freedom in a system. A degree of freedom is any feature of the system that can vary over time. A human arm, for example, can move in seven independent directions and so has seven degrees of freedom: three at the shoulder, which can move up and down, forward and backward, and rotate; one at the elbow, which can move toward and away from the upper arm; one at the radio-ulnar joint, which can rotate; and two at the wrist, which can move up and down and side to side. This suggests that to control the movement of a human arm, the brain must closely control seven independent degrees of freedom, a daunting task. In actual human movement, however, there is a task-dependent coupling of several of these degrees of freedom, such that one degree of freedom is connected to or dependent on others; that is, they form synergies.[3] When degrees of freedom are coupled, the number of dimensions along which the arm's movement can vary is reduced. This is dimensional compression. Dimensional compression is significant because it simplifies control and improves a system's functional stability. Synergies exhibit dimensional compression. We find dimensional compression, and in so doing, show that something is a synergy, using principal component analysis. Principal component analysis is a widely used statistical technique that identifies covariation within high-dimensional data sets and remaps the data (taking the covariation into account) into a space whose axes (principal components) represent the data set's primary dimensions of variation.[4] Those dimensions, termed the principal components, typically do not relate directly to the original measurement dimensions. If the original variables are in fact coupled, principal component analysis yields a dimension reduction—fewer principal components are required to account for most of the variance in the data set than the number of original variables. This is a measure of dimensional compression.

Here is an example. Place your upper arm on a tabletop. The tabletop eliminates three degrees of freedom by eliminating shoulder movements, leaving four degrees of freedom at the elbow and wrist. Now rapidly flex and extend your elbow. This movement creates torque at your wrist joint, which should cause your hand to flap. But it doesn't flap. This is because a wrist-elbow synergy makes it so that the degrees of freedom at the wrist and the elbow constrain one another, so that the wrist and elbow movements are highly correlated. Principal component analysis here would allow us to see this in the reduction of the dimensions required to account for the movements of the arm from four to three.

Getting back to Nalepka's experiments, for the tabletop two-person video game data, we did a principal component analysis of the movements of successful pairs of players. Their movements varied in vertical and horizontal directions on the tabletop, so each player contributed two dimensions, yielding four total degrees of freedom. Principal component analysis showed that the first principal component, compressing the number of dimensions of the system from four to three, accounted for more than half of the variability. That is, there was one virtual dimension, an amalgam of the measured movements of the players, that accounted for half of the data. This variable is a relationship among the movements of the players while playing the game. The players formed an interpersonal synergy, a unified system of parts that spanned both players, along with the video game they played.

## CHAPTER 5: JAZZ PERFORMANCE

To analyze the movements of the jazz musicians, Ashley Walton and colleagues use cross wavelets analysis. Richard Schmidt and colleagues have demonstrated its use in understanding the movement coordination that occurs between actors during joke telling and dancing.[5] The advantage of cross wavelet analysis is its ability to reveal common periodicities in behavioral coordination at nested timescales, detecting local microscale structures (e.g., note or bar) within global macroscale patterns (e.g., chorus or piece).

Cross wavelet analysis assesses coordination between two time series through examination of the coupling strength and relative phase of the coordination that occurs between participants across multiple timescales.[6] As with detrended fluctuation analysis, described above, the strength of coupling and the relative phase angle between two time series is assessed for shorter, half second and second-to-second timescales, as well as at longer four-, eight-, twelve-, and sixteen-second timescales.[7] We used cross wavelet analysis to investigate the coordination between the lateral movements of the forearms and head movements of two piano players. For each of the different timescales, the average distribution of relative phase angles that occurred between the musicians' movement time series is calculated. Recall that relative phase is a collective variable, embodied not in either musician but in the coordination between them.

When the musicians who were asked not to improvise but to play along with the ostinato backing track, there was very strong coupling and nearly perfectly coordinated hand movements at four-second intervals, the length of the repeating ostinato phrase. There was also strong, intermittent coupling at shorter timescales corresponding to the playing of the notes that make up the ostinato phrase. This is what near-perfect coordination looks like when analyzed using cross wavelet spectral analysis. In contrast, when musicians improvised over the jazz standard backing track, their head movements are more highly coordinated than their hand movements, at a much faster timescale, corresponding to the rhythm of backing track. While the hand movements may be considered more functionally relevant for musical production, velocity of head movements have been found to play a large role in performance expressivity.[8] Compare head movements to the actual improvised melodies the musicians are playing; the layout of a piano means that the higher notes, those typically used to play melodies, are toward the right side of the keyboard and played with the right hand. In contrast to the highly coordinated head movements, there is almost no strong coupling between the musicians' melody-playing hands. Unlike the head movements, the infrequent, intermittent coupling in the melody-playing hands is typically out of phase.

### CHAPTER 5: COMPLEXITY MATCHING

Drew Abney and colleagues looked for nested structure using Allan factor analysis. As with DFA and wavelet analysis, Allan factor analysis divides the full trial into windows of varying sizes and counts the vocalizations and movements in each window. The average difference in the number of vocalizations and movements between neighboring windows is called Allan factor variance and is the basis for measuring structure. The average difference increases when events (vocalizations and/or movements) group together more within particular windows but less so in neighboring windows. Abney and colleagues calculated the average difference for each timescale, which produces a measure of structure as a function of timescale. Structure is nested when structured events in short timescale windows combine to form larger structures in longer timescale windows, which combine to form still larger structures in longer timescale windows, and so on, until the window encompasses the whole of the trial. The degree of nesting in temporal structures is measured by changes in Allan factor variance as a function of timescale.[9]

# META-APPENDIX:
# THE CONTROVERSY OVER 1/*F* NOISE

**IT HAS LONG BEEN KNOWN** that self-organized critical systems exhibit a specific, fractal variety of fluctuation called $1/f$ noise or pink noise.[1] $1/f$ noise or $1/f$ scaling or pink noise—these terms are used synonymously in the literature—is a kind of not-quite-random, correlated noise, halfway between genuine randomness (white noise) and a drunkard's walk, in which each behavior is constrained by the prior one (brown noise). $1/f$ noise is often described as a fractal structure in a time series, in which the variability at a short timescale is correlated with variability at a longer timescale. (This is the long memory of interaction-dominant systems described in chapter 5.) Interaction-dominant dynamics predicts that this $1/f$ noise would be present. As discussed in chapter 5, the fluctuations in an interaction-dominant system percolate through the system over time, leading to the kind of correlated structure to variability that is $1/f$ noise. Here is the way to infer that cognitive and neural systems are synergies: exhibiting $1/f$ noise is evidence that a system is interaction dominant; interaction-dominant systems are synergies. The mounting evidence that $1/f$ noise is ubiquitous in human physiological systems, behavior, and neural activity is also evidence that human physiological, cognitive, and neural systems are interaction dominant, which in turn is evidence that they are synergies. However, the inference from $1/f$ noise to interaction dominance has been controversial. In this appendix, I trace some of this controversy. To begin, I describe American philosopher Charles Saunders Peirce's views on abduction in the sciences. I then show how abductive reasoning is used by both sides of the controversy over complex systems explanations in the cognitive sciences. I point to some methodological advances and empirical evidence that bear on the controversy. I will ultimately argue that the complex systems variety of dynamical cognitive science is alive and well in the 2020s and that this does not imply that other approaches are moribund.

## (FRANKFURT ON) PEIRCE ON ABDUCTION

Peirce's work on abductive inferences in the sciences was brought into the cognitive sciences by Jerry Fodor. Fodor argued that abduction shows that cognition cannot be (only) computation. Although widely disputed, this is a rare point of agreement between Fodor and those outside the mainstream of cognitive science, like the phenomenological philosopher John Haugeland and the ecological psychologist Michael Turvey.[2] Cognition, these disparate thinkers agree, can be feasibly understood as computation only if it involves informational encapsulation, that is, modules. Abduction seems to defy informational encapsulation and involve creative synthesis. What then is abduction? Peirce's own writing on this subject is notoriously convoluted and jargon laden, so I will draw on a classical presentation of Peirce's work on scientific reasoning by Harry Frankfurt.[3] According to Peirce, according to Frankfurt, there are three basic kinds of scientific inference: abduction, deduction, and induction. In abduction, one hypothesizes an explanation for a surprising observation. It has the following form:

1. The surprising fact, F, is observed.
2. But if explanation E were true, F would be a matter of course.
3. Hence, there is reason to suspect that E is true.

This kind of inference to the best explanation is a crucial part of both science and everyday cognition. While preparing to cook dinner, we notice that the tofu is missing from the refrigerator and try to explain that fact. It would be unsurprising for the tofu to be gone if some member of our family had made miso soup for lunch today. We reasonably come to suspect that someone made miso soup for lunch. In a more scientific context, we notice that dolphins, unlike most other sea creatures, frequently come to the surface and seem to breathe air. One possible explanation for this is that dolphins, despite their appearance and life in the ocean, are not fish. Abduction involves creativity and bringing to bear arbitrary pieces of information from across subject areas. Abductive inference will be central to the disputes described below.

Peirce also identified two additional, more familiar types of scientific and everyday inferences, which follow abduction in attempting to gather scientific evidence. First, there is deduction, which takes the following form.

1. Derive further testable consequences C of explanation E.
2. Test those consequences C.
3. Observing that C is evidence for E.

Deduction will be familiar to those who have studied the philosophy of science—it is often called the "deductive-nomological model of explanation."[4]

The third type of scientific reasoning Peirce identifies is induction. In induction, one iterates the deductive process and decides on the correctness of explanatory hypothesis E, based on some criterion for the required number of observations of C to make it a matter of course. For Peirce, the scientific process starts with abduction, in which we generate an explanation for some phenomenon, moves on to deduction, in which we attempt to generate evidence that the explanation is the correct one, and moves finally to induction, where we attempt to decide whether the evidence we have gathered is sufficient to demonstrate that the explanation is correct. In the rest of this appendix, I will show this pattern—abduction to deduction to induction—at play in dynamical cognitive science.

## ABDUCTION AND DEDUCTION IN
## DYNAMICAL COGNITIVE SCIENCE

In a 2003 paper,[5] Guy van Orden, Jay Holden, and Michael Turvey offer a version of Peirce's abductive inference based on the "surprising fact" of the ubiquity of $1/f$ noise in human behavior:

1. Ubiquitous $1/f$ noise in human behavior is observed.
2. But if the human cognitive system is interaction dominant, $1/f$ noise in human behavior would be a matter of course.
3. Hence, there is reason to suspect that the human cognitive system is an interaction-dominant system.

Here the proposed explanation for the ubiquity of $1/f$ noise is that the human cognitive system is interaction dominant. Van Orden et al. then engage in a Peircean deduction:

1. A consequence of the interaction dominance of the human cognitive system is that we should also see $1/f$ noise in cognition.
2. Test this consequence by looking for $1/f$ noise in human cognition.
3. Observing $1/f$ noise in human cognition is evidence that human cognitive systems are interaction dominant.

Van Orden et al. did two experiments to look for $1/f$ noise in human cognition. In one, participants were asked to say /ta/ in response to a visual cue for 45 practice trials and then 1,100 test trials; in the second, participants were asked to say words that appeared on a monitor for 45 practice trials and 1,100 experimental trials. The data in both cases was a time series of reaction times. First, it might seem strange that there were so many trials per participant; this was necessary because the time series analyses used to test for $1/f$ noise generally require comparisons of three orders of magnitude. Second, unlike the finger wagging that was modeled by Hermann Haken, Scott Kelso, and Herbert Bunz, responding to a stimulus and reading words are straightforwardly cognitive activities and are tasks commonly used in experiments across the cognitive sciences. The findings in both van Orden et al.'s experiments were that participant behavior exhibited $1/f$ noise, providing evidence that human cognitive systems are interaction dominant.

That the human cognitive system is interaction dominant is a very strong claim. For one, it supports the claims discussed above by Fodor, Turvey, and Haugeland that cognition could not be (just) computation. For Fodor, the claim is that peripheral cognition might be modular and computational, but what he calls central cognition cannot be. For Turvey and Haugeland, the claim is that none of cognition is well understood as modular or computational.

### AN ALTERNATIVE ABDUCTION AND DEDUCTION

One of the things we teach introductory psychology and cognitive science students is that an appropriate response to experimental evidence is to propose an alternative hypothesis that might explain the same data. This was exactly the response to van Orden et al. from E. J. Wagenmakers, Simon Farrell, and Roger Radcliff.[6] Van Orden et al. explained the $1/f$ noise in cognition appears because cognitive systems are interaction dominant. Wagenmakers et al. responded that we might see

$1/f$ noise in a modular component-dynamic cognitive system because when the modules generate white noise, adding the white noise of each module together, the system might produce $1/f$ noise at its output. Here is their abduction:

1. Ubiquitous $1/f$ noise in human behavior is observed.
2. But if the human cognitive system is component dominant and each component contributes uncorrelated white noise to behavior, $1/f$ noise in human behavior would be a matter of course.
3. Hence, there is reason to suspect that the human cognitive system is a component-dominant system, with each component contributing uncorrelated white noise to behavior.

See figure A2.1. As evidence for this hypothesis the deductive step in Peirce's schema, Wagenmakers et al. show via simulations

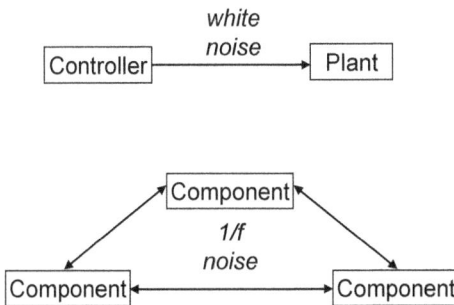

**FIGURE A2.1** The differences between component-dominant and interaction-dominant dynamics. Image created by author.

that a component-dominant system in which each component contributes white noise can generate data with $1/f$ noise.

Van Orden et al. respond to Wagenmakers et al. by arguing that their simulations are gerrymandered and implausible. Their "newly proposed encapsulated mechanisms lack physical or psychological motivation before the fact and require that one posits implausible coincidences after the fact."[7] This may indeed be the case, but it does not keep the Wagenmakers et al. abduction from presenting a live alternative hypothesis. And an alternative hypothesis in the sciences is not the same as a refutation; it is an invitation to additional research.

## ADDITIONAL RESEARCH

As an attempt to mediate this dispute, Espen Ihlen and Beatrix Vereiken reexamine both the van Orden et al. human data and the Wagenmakers et al. simulation data using a wavelet-based multifractal analysis.[8] What they found was that, although both the human and simulation data do exhibit $1/f$ noise, only the human data exhibits multifractality. The simulation data is monofractal in that it exhibits the same degree of $1/f$ noise across all timescales. The human data is multifractal in that it exhibits different degrees of $1/f$ noise at different timescales. The upshot of this work is that $1/f$ noise is not sufficient for claiming that a system is interaction dominant. Instead, the standard ought to be multifractality. This methodological innovation changes the requirements for arguing that the human cognitive system is interaction dominant, and so not modular or computational, from exhibiting $1/f$ noise to exhibiting multifractality.

In a related development, dynamical cognitive scientists have explored the relationship between multifractality and complexity matching. In complexity matching, two systems maximize the information transfer between them by coordinating the degree of their $1/f$ noise across multiple temporal scales.[9] Complexity matching is a consequence of multifractality; for two interacting cognitive systems to be able to match their degree of $1/f$ noise, they must be capable of organizing their behavior to control their degree of $1/f$ noise over time in concert with their partner.[10] This in turn requires that cognitive systems be capable of moving from one degree of $1/f$ noise to another in their activity; that is, it requires that their performance is multifractal. So complexity matching is evidence of multifractality, which makes it also evidence of interaction dominance.

This leads to a revision of the van Orden et al. abduction:

1. Multifractality is observed in human behavior.
2. But if the human cognitive system is interaction dominant, multifractality in human behavior would be a matter of course.
3. Hence, there is reason to suspect that human cognitive system is an interaction-dominant system.

Lillian Rigoli, Daniel Holman, Michael Spivey, and Christopher Kello use complexity matching to develop what might be called an *experimentum crucis*, explicitly designed for deciding between the Van Orden et al. and Wagenmakers et al. abductions. They divided participants into two groups and asked them to tap a keyboard spacebar along with a metronome, with one group tapping on beat and the other in syncopation for 1,100 taps in an approximately fifteen-minute trial. They measured time series of pupil dilations, heartbeat intervals, keypress durations, and keypress deviations. The van Orden et al. abduction and the Wagenmakers et al. abduction each imply different results. If human cognition is interaction dominant, we should expect each time series to separately exhibit $1/f$ noise and for the physiological measures (pupil dilation, heartbeat interval) and the behavioral measures (keypress duration and interval), respectively, to match one another's complexity. If human cognition is component dominant, we should expect neither $1/f$ noise in each measurement nor complexity matching among them. The results clearly vindicated the van Orden et al. abduction. Each of the time series measured (pupil dilations, heartbeat intervals, keypress durations, and keypress deviations) exhibited $1/f$ noise. As we have seen, this is insufficient to guarantee that the systems in question are interaction dominant. There were significant correlations, however, between the physiological measures of $1/f$ noise and between the behavioral measures of $1/f$ noise. This complexity matching is indicative of multifractality in each of these systems; that is, it indicates that they are interaction dominant.[11]

### DEDUCTION IN DYNAMICAL COGNITIVE SCIENCE

According to Peirce, abduction in science ought to be followed by deduction and eventually induction. The ongoing deductive phase in complex-systems dynamical cognitive science takes the following form:

1. A consequence of the interaction dominance of the human cognitive system is that we should see complexity matching and multifractality in cognitive phenomena.
2. Test this consequence by looking for complexity matching and multifractality in human cognition.
3. Observing complexity matching and multifractality in human cognition is evidence that human cognitive systems are interaction dominant.

The hypothesis that human cognition is interaction dominant has been tested repeatedly over the last decade, and there are literally hundreds of published findings demonstrating complexity matching and multifractality in aspects of human cognition. A small, idiosyncratic selection of these have been described earlier in chapter 5. More generally, we see that multifractality and complexity matching are ubiquitous in physiology, neural activity at every scale, human and animal movement, cognitive tasks, human-tool systems, and human social systems, suggesting that each scale is interaction dominant.

This goes beyond the rejection of modularity that we saw in Fodor, Haugeland, and Turvey to a rethinking of the nature and extent of cognitive systems. Recall from chapter 5 that interaction dominance is indicative of well-functioning physiological systems. This suggests that processes at each scale are as densely interconnected as the components of a human heart. Because these examples of interaction-dominant systems (among many others[12]) also span human-tool systems and groups of humans, there is good reason to think that extended and multiperson cognitive systems are as real and robust as individual human cognizers. Yet the cognitive sciences focus almost exclusively on the smallest of these interaction-dominant systems (neurons, brain areas, whole brains) and neglect the larger-scale, beyond-the-individual ones.[13] From a complex-systems dynamical point of view, there is nothing special about biological or psychological individuals.

## CODA AND EXERCISES FOR READERS

With so many successful deductions providing evidence for the interaction dominance of human cognitive systems and even an *experimentum crucis*, we should wonder whether we have arrived at Peirce's induction phase. In induction, one weighs the evidence adduced in favor of an explanation of an observed phenomenon with an eye toward determining whether that evidence is sufficient to make the explanation part of the accepted-but-still-revisable scientific canon. What the criteria are for this is hardly clear. I leave the development of such criteria and the determination of whether the interaction dominance of human cognitive systems is the correct explanation for the ubiquitous multifractality and complexity matching in human behavior as exercises for readers.

For the sake of argument, suppose that the hypothesis of the interaction dominance of human cognition is in Peirce's induction phase. This would imply that it should appear in textbooks, at the very least, as should discussion of extended human-tool and multiperson cognitive systems. And now it does, thanks to Michael Spivey's additions to the third edition of the textbook *Cognitive Science: An Introduction to the Study of Mind.*[14] What it does not imply is that other hypotheses about the nature of cognition are false or inadequate. What this suggests is that, at least in its current state, cognitive science demands a plurality of explanations.[15] Some aspects of cognition are demonstrably interaction dominant; others might best be explained in terms of neural network dynamics; still others might best be explained in terms of computation. For the time being, despite the progress in complex systems explanations described in this appendix and in chapters 4 and 5, dynamical cognitive science remains a big tent, as does cognitive science as a whole.

# NOTES

## PREFACE

1. Noam Chomsky, *What Kind of Creatures Are We?* (Columbia University Press, 2015).
2. Daniel C. Dennett, *The Intentional Stance* (MIT Press, 1987).
3. Richard Hernnstein and Charles Murray, *The Bell Curve* (Simon and Schuster, 1994).
4. Sally Haslanger, "Gender and Race: (What) Are They? (What) Do We Want Them to Be?," *Nous* 34 (2000): 31–55.
5. *The World's End*, dir. Edgar Wright (2013; Working Title Films).

## 1. OTHER MINDS

1. Alan Gabbey, "Reflections on the Other Minds Problem: Descartes and Others," in *Sceptics, Millenarians and Jews*, ed. David S. Katz and Jonathan I. Israel (Brill, 1990).
2. Thanks to Zvi Biener for advice on this.
3. Moira Gatens, *Feminism and Philosophy: Perspectives on Difference and Equality* (Indiana University Press, 1991).
4. John Locke, *An Essay Concerning Human Understanding*, 1690, *An Essay*, I, 3, 25, https://www.gutenberg.org/files/10615/10615-h/10615-h.htm.

5. Locke, *An Essay*, 2, 27, 17.
6. Locke, *An Essay*, 2, 27, 26.
7. John Locke, *Two Treatises of Government*, 1690, 2, 4, https://www.gutenberg.org/files/7370/7370-h/7370-h.htm.
8. Locke, *Two Treatises*, 5, 8.
9. Martin Heidegger, *Being and Time*, trans. John Macquarrie and Edward Robinson (HarperCollins, 1962).
10. Heidegger, *Being and Time*, 155.
11. Heidegger, *Being and Time*, 154.
12. Maurice Merleau-Ponty, *Phenomenology of Perception*, trans. Donald Landes (Routledge, 2012), 273.
13. Andy Clark, *Being There: Putting Brain, Body, and World Together Again* (MIT Press, 1997); Andy Clark and David Chalmers, "The Extended Mind," *Analysis* 58, no. 1 (1998): 7–19.
14. This set of distinctions is drawn from Hanne De Jaegher, Ezequiel Di Paolo, and Shaun Gallagher, "Can Social Interaction Constitute Social Cognition?," *Trends in Cognitive Sciences* 14, no. 10 (2010): 441–47.
15. Merleau-Ponty, *Phenomenology of Perception*, 413.
16. Maurice Merleau-Ponty, *Résumés de Cours: Collège de France, 1952–1960* (Gallimard, 1968). This is my translation. The original French: "Nous n'avons plus à comprendre comment un Pour Soi peut en penser un autre à partir de sa solitude absolue ou peut penser un monde préconstitué au moment même où il le constitue : *l'inhérence du soi au monde ou du monde au soi, du soi à l'autre et de l'autre au soi*, ce que Husserl appelle l'Ineinander, est silencieusement inscrit dans une expérience intégrale, ces incompossibles sont composés par elle, et la philosophie devient la tentative, par-delà la logique et le vocabulaire donnés, de décrire cet univers de paradoxes vivants." (italics added). The published English translation is shockingly bad. The phrase I have italicized above is rendered "the inherence of the self-in-the-world or of the world-in-the-self" (108), with the second half of the phrase, concerning the self and other, simply ignored.
17. For Merleau-Ponty experts: Some Merleau-Ponty scholars view his posthumously published work to be a departure from his earlier phenomenology. I follow Daly and Müller in thinking that Merleau-Ponty's thinking is continuous across his career on this matter. Krueger finds what Merleau-Ponty later calls the *Ineinander* in his Sorbonne lectures from 1948. Anya Daly, *Merleau-Ponty and the Ethics of Intersubjectivity*

(Palgrave Macmillan, 2016); Robin Müller, "The Logic of the Chiasm in Merleau-Ponty's Early Philosophy," *Ergo* 4 (2017); Joel Krueger, "Merleau-Ponty on Shared Emotions and the Joint Ownership Thesis," *Continental Philosophy Review* 46 (2013): 509–31.

## 2. THE EMBODIED MIND

1. See James J. Gibson, *The Ecological Approach to Visual Perception* (Houghton-Mifflin, 1979); Anthony Chemero, *Radical Embodied Cognitive Science* (MIT, 2009). Stephan Käufer, and Anthony Chemero, *Phenomenology: An Introduction*, 2nd ed. (Wiley, 2021). This brief description and what follows is based on human perception and action. Louise Barrett, *Beyond the Brain: How Body and Environment Shape Animal and Human Minds* (Princeton University Press, 2011) provides an excellent description of the ecological approach in the context of nonhuman perception and action.
2. Gibson, *The Ecological Approach to Visual Perception*, 127.
3. Gibson, *The Ecological Approach to Visual Perception*, 129.
4. For history of the development of Gibson's ecological approach, see Harry Heft, *Ecological Psychology in Context: James Gibson, Roger Barker, and the Legacy of William James's Radical Empiricism* (Psychology Press, 2001): Lorena Lobo, Manuel Heras-Escribano, and David Travieso, "The History and Philosophy of Ecological Psychology," *Frontiers in Psychology* 9 (2018): 2228; Jelle Bruineberg, Rob Withagen, and Ludger Van Dijk, "Productive Pluralism: The Coming of Age of Ecological Psychology," *Psychological Review 131*, no. 4: 993–1006; Miguel Segundo-Ortin and Vicente Raja, *Ecological Psychology* (Cambridge University Press, 2024).
5. Käufer and Chemero, *Phenomenology*.
6. Francisco Varela, Evan Thompson, and Eleanor Rosch, *The Embodied Mind* (MIT Press, 1991).
7. Rodney A. Brooks, "Intelligence Without Representation," *Artificial Intelligence* 47, no. 1–3 (1991): 139–59.
8. Evan Thompson, *Mind in Life* (Harvard University Press, 2007). Somewhat different understandings of enactivism are developed in Kevin O'Regan and Alva Noë, on one hand, and Dan Hutto and Erik Myin, on the other. Rather than starting with a conception of life, O'Regan and Noë's enactive approach focuses on sensorimotor contingencies.

Experience is possible, they argue, because we have implicit knowledge of connections about how we move and how the world looks. Hutto and Myin come from the analytic tradition and marshal a series of effective arguments against the idea that basic cognition involves internal mental representations. Each of these views ends up in roughly the same spot, but the path to that spot is different in each case. Kevin O'Regan, and Alva Noë, "A Sensorimotor Account of Vision and Visual Consciousness," *Behavioral and Brain Sciences* 24 (2001): 939–73; Alva Noë, *Strange Tools* (Hill and Wang, 2015); Alva Noë, *The Entanglement* (Princeton University Press, 2023); Daniel Hutto, and Erik Myin, *Radicalizing Enactivism: Basic Minds without Content* (MIT Press, 2013).

9. Francisco J. Varela, *Principles of Biological Autonomy* (North-Holland, 1979); Thompson *Mind and Life*; Evan Thompson and Mog Stapleton, "Making Sense of Sense-Making: Reflections on Enactive and Extended Mind Theories," *Topoi* 28 (2009): 23–30.

10. Hanne De Jaegher and Ezequiel Di Paolo, "Participatory Sense-Making: An Enactive Approach to Social Cognition," *Phenomenology and the Cognitive Sciences* 6 (2007): 485–507. See also Glenda Satne, "Understanding Others by Doing Things Together: An Enactive Account," *Synthese* 198, no. Suppl 1 (2021): 507–28.

11. Thomas Fuchs, *Ecology of the Brain: The Phenomenology and Biology of the Embodied Mind* (Oxford University Press, 2018): Thomas Fuchs and Hanne De Jaegher, "Enactive Intersubjectivity: Participatory Sense-Making and Mutual Incorporation," *Phenomenology and the Cognitive Sciences* 8 (2009): 465–86.

12. Participatory sense-making is closely related to what Alva Noë has called "coordinated activities." The empirical results developed in chapter 5 in support of the existence of participatory sensemaking, sensorimotor empathy, and the intertwined self also support Noë's philosophical ideas. See Noë, *Strange Tools*; Noë, *The Entanglement*.

13. Beginning with Anthony Chemero, *Radical Embodied Cognitive Science* (MIT Press, 2009).

14. For example, Erik Rietveld and Julian Kiverstein, "A Rich Landscape of Affordances," *Ecological Psychology* 26 (2014): 325–52; Ezequiel A. Di Paolo, Manuel Heras-Escribano, Anthony Chemero, and Matthew McGann, eds., *Enaction and Ecological Psychology: Convergences and Complementarities* (Frontiers, 2021).

15. Varela et al., *The Embodied Mind*, 204.

16. R. Swenson, "Autocatakinetics, Yes-Autopoiesis, No: Steps Towards a Unified Theory of Evolutionary Ordering," *International Journal of General Systems* 21, no. 2 (1992): 207–28; R. Swenson and M. T. Turvey, "Thermodynamic Reasons for Perception-Action Cycles," *Ecological Psychology* 3, no. 4 (1991): 317–48.

17. Rob Withagen, Harjo J. De Poel, Duarte Araújo, and Gert-Jan Pepping, "Affordances Can Invite Behavior: Reconsidering the Relationship Between Affordances and Agency," *New Ideas in Psychology* 30, no. 2 (2012): 250–58; Rob Withagen, *Affective Gibsonian Psychology* (Taylor & Francis, 2022); Rob Withagen, "The Field of Invitations," *Ecological Psychology* 35, no. 3 (2023): 102–15.

18. Mog Sapleton, "Enactivism Embraces Ecological Psychology," *Constructivist Foundations* 11 (2016): 325–27; M. McGann, "Enacting a Social Ecology: Radically Embodied Intersubjectivity," *Frontiers in Psychology* 5 (2014): 1321; J. McKinney, S. V. Steffensen, and A. Chemero, "Practice, Enactivism, and Ecological Psychology," *Adaptive Behavior* 31, no. 2 (2023): 143–49.

19. Shaun Gallagher, *Enactivist Interventions: Rethinking the Mind* (Oxford University Press, 2017).

20. Ezequiel A. Di Paolo, Thomas Buhrmann, and Xabier Barandiaran, *Sensorimotor Life: An Enactive Proposal* (Oxford University Press, 2017).

21. H. Heft, "Ecological Psychology and Enaction Theory: Divergent Groundings," *Frontiers in Psychology* 11 (2020): 991.

22. Rietveld and Kiverstein, "A Rich Landscape of Affordances"; Ludger Van Dijk and Erik Rietveld, "Foregrounding Sociomaterial Practice in Our Understanding of Affordances: The Skilled Intentionality Framework," *Frontiers in Psychology* 7 (2017); Jelle Bruineberg, Anthony Chemero, and Erik Rietveld, "General Ecological Information Supports Engagement with Affordances for 'Higher' Cognition," *Synthese* 196 (2019): 5231–51.

23. Rietveld and Kiverstein, "A Rich Landscape of Affordances," 328–29.

24. Edward Baggs, "All Affordances Are Social: Foundations of a Gibsonian Social Ontology," *Ecological Psychology* 33, no. 3–4 (2021): 257–78; Shaun Gallagher and Taylor G. Ransom, "Artifacting Minds: Material Engagement Theory and Joint Action," in *Embodiment in Evolution and Culture*, ed. G. Etzelmüller and C. Tewes (Mohr Siebeck, 2016); Miguel Segundo-Ortin and Glenda Satne, "Sharing Attention, Sharing Affordances: From Dyadic Interaction to Collective Information," in *Access and Mediation*, ed. M. Wehrle, D. D'Angelo, and E. Solomonova (DeGruyter, 2022).

25. Nick Brancazio, "Being Perceived and Being 'Seen': Interpersonal Affordances, Agency, and Selfhood," *Frontiers in Psychology* 11 (2020): 1750, 8.
26. Frantz Fanon, *Black Skin, White Masks*, trans. Charles Lam Markmann (Grove Press, 1967), 82.
27. See M. Heras-Escribano, *The Philosophy of Affordances*. (Palgrave Macmillan, 2019).

### 3. THE INTERTWINED SELF

1. Edward Evarts, "Sherrington's Concept of Proprioception," *Trends in Neurosciences* 4 (1981): 44–46; J. Pearce, "Sir Charles Scott Sherrington (1857–1952) and the Synapse."*Journal of Neurology, Neurosurgery & Psychiatry*, 75 (2004): 544.
2. Ernst Mach, *The Analysis of Sensations and the Relation of the Physical to the Psychical.*, trans. C. M. Williams (Open Court, 1914), originally published as *Beiträge zur Analyse der Empfindungen* (Gustav Fischer, 1886).
3. James J. Gibson, *The Ecological Approach to Visual Perception* (Houghton Mifflin, 1979), 115
4. David Lee and Erik Aronson, "Visual Proprioceptive Control of Standing in Human Infants," *Perception & Psychophysics* 15 (1974): 529–32.
5. David N. Lee and J. R. Lishman, "Visual Proprioceptive Control of Stance,"*Journal of Human Movement Studies* (1975): 87–95.
6. Ulrich Neisser, "Five Kinds of Self-Knowledge," *Philosophical Psychology* 1, no. 1 (1988): 35–59; Ulrich Neisser, "The Self Perceived," in *The Perceived Self: Ecological and Interpersonal Sources of Self-Knowledge*, vol. 5, ed. Ulric Neisser (Cambridge University Press, 1993), 3–21.
7. Neisser, "The Self Perceived," 204.
8. Neisser. "The Self Perceived," 10.
9. Alva Noë uses this example effectively. See Alva Noë, *Strange Tools* (Hill and Wang, 2015).
10. Shaun Gallagher, "Philosophical Conceptions of the Self: Implications for Cognitive Science," *Trends in Cognitive Sciences* 4, no. 1 (2000), 15.
11. Shaun Gallagher, and Dan Zahavi. *The Phenomenological Mind* (Routledge, 2020).
12. Sanneke de Haan and Leon de Bruin, "Reconstructing the Minimal Self, or How to Make Sense of Agency and Ownership," *Phenomenology and the Cognitive Sciences* 9 (2010): 373–96.

13. Daniel Zahavi, *Self and Other: Exploring Subjectivity, Empathy, and Shame* (Oxford University Press, 2014).

14. Daniel Dennett, *Consciousness Explained* (Little Brown, 1991).

15. Shaun Gallagher and Daniel Hutto, "Understanding Others Through Primary Interaction and Narrative Practice," in *The Shared Mind: Perspectives on Intersubjectivity*, ed. Jordan Zlatev, Timothy P. Racine, Chris Sinha, and Esa Itkonen. John Benjamins (John Benjamins, 2007), 12–38; D. D. Hutto, *Folk Psychological Narratives: The Sociocultural Basis of Understanding Reasons* (MIT Press, 2012).

16. Daniel Zahavi, "Empathy, Embodiment and Interpersonal Understanding: From Lipps to Schutz," *Inquiry* 53 (2010): 285–306; Zahavi, *Self and Other*.

17. Shaun Gallagher, "A Pattern Theory of Self," *Frontiers in Human Neuroscience* 7 (2013): 443.

18. Zahavi, "Empathy, Embodiment, and Interpersonal Understanding"; Zahavi, *Self and Other*.

19. There are other approaches that take the self to be embodied and social. For an overview, see R. Heersmink, "Varieties of the Extended Self," *Consciousness and Cognition* 85 (2020): 103001.

20. Sanneke de Haan, "The Minimal Self Is a Social Self," in *The Embodied Self: Dimensions, Coherence, and Disorders*, ed. Thomas Fuchs, Heribert Sattel, and Peter Henningsen (Schattauer, 2010), 12.

21. M. Merleau-Ponty, *The Visible and the Invisible* (Northwestern University Press, 1964).

22. Vasudevi Reddy, "On Being the Object of Attention: Implications for Self–Other Consciousness," *Trends in Cognitive Sciences* 7, no. 9 (2003): 397–402. Vasudevi Reddy, *How Infants Know Minds* (Harvard University Press, 2008).

23. Along with several colleagues. See Anna Ciaunica, "The 'Meeting of Bodies': Empathy and Basic Forms of Shared Experiences," *Topoi* 38, no. 1 (2019): 185–95; Anna Ciaunica, Axel Constant, Hubert Preissl, and Katerina Fotopoulou, "The First Prior: From Co-Embodiment to Co-Homeostasis in Early Life," *Consciousness and Cognition* 91 (2021): 103117.

24. Alejandra Martínez Quintero and Hanne De Jaegher. "Pregnant Agencies: Movement and Participation in Maternal–Fetal Interactions," *Frontiers in Psychology* 11 (2020): 1977.

25. Miriam Kyselo, "The Body Social: An Enactive Approach to the Self," *Frontiers in Psychology* 5 (2014): 986; Miriam Kyselo, "The Minimal Self Needs a Social Update," *Philosophical Psychology* 29, no. 7 (2016): 1057–65.

26. Miriam Kyselo, "The Minimal Self Needs a Social Update," *Philosophical Psychology* 29, no. 7 (2016): 1057–65.
27. Joel Krueger, "Merleau-Ponty on Shared Emotions and the Joint Ownership Thesis," *Continental Philosophy Review* 46 (2013): 509–31.
28. Anthony Chemero, "Sensorimotor Empathy," *Journal of Consciousness Studies* 23, no. 5–6 (2016): 138–52.
29. Heidi Lene Maibom, ed., *Empathy and Morality* (Academic, 2014).
30. Giovanna Colombetti, *The Feeling Body: Affective Science Meets the Enactive Mind* (MIT Press, 2014).
31. Shaun Gallagher, *How the Body Shapes the Mind* (Clarendon Press, 2005); Shaun Gallagher, "A Well-Trodden Path: From Phenomenology to Enactivism," *Filosofisk Supplement* 3 (2018): 42–47.
32. Joel Krueger, "Direct Social Perception," in *The Oxford Handbook of 4E Cognition*, ed. Albert Newen, Leon de Bruin, and Shaun Gallagher (Oxford University Press, 2018).

## 4. RADICAL EMBODIED COGNITIVE SCIENCE

1. Jerry Fodor, *Psychosemantics: The Problem of Meaning in the Philosophy of Mind* (MIT Press, 1987). Thanks to Peter-Langland-Hassan for helping me find the passage.
2. Jerry A. Fodor, "Methodological Solipsism Considered as a Research Strategy in Cognitive Psychology," *Behavioral and Brain Sciences* 3, no. 1 (1980): 63–73.
3. Andy Clark, *Being There: Putting Brain, Body, and World Together Again* (MIT Press, 1997).
4. Martin Heidegger, *Being and Time*, trans. John Macquarrie and Edward Robinson (HarperCollins, 1962), 89.
5. In *Radical Embodied Cognitive Science*, I presented radical embodied cognitive science primarily in terms of skepticism about internal mental representations. Dan Hutto and Erik Myin have developed a series of arguments concerning mental representation in basic cognition. My book, in their terms, was a "don't need" argument against mental representations: I proposed explaining cognition without referring to mental representations. Hutto and Myin develop "can't have" arguments, showing convincingly that the very idea of internal mental representations is untenable. See Daniel D. Hutto and Erik Myin, *Evolving Enactivism: Basic Minds Meet Content* (MIT Press, 2017).

6. James J. Gibson, *Ecological Approach to Visual* Perception (Houghton-Mifflin, 1979); Peter N. Kugler, J. A. Scott Kelso, and Michael T. Turvey, "1 on the Concept of Coordinative Structures as Dissipative Structures: I. Theoretical Lines of Convergence," in *Advances in Psychology*, vol. 1, ed. George E. Stelmach and Jean Requin (North-Holland, 1980).

7. Randall D. Beer, "A Dynamical Systems Perspective on Agent-Environment Interaction," *Artificial Intelligence* 72, no. 1–2 (1995): 173–215.

8. Not only in the cognitive sciences but also in the rehabilitation sciences and in design. See Chris Baber, *Embodying Design: An Applied Science of Radical Embodied Cognition* (MIT Press, 2022): Simon Penny, *Making Sense: Cognition, Computing, Art and Embodiment* (MIT Press, 2018); Paula Silva, Adam Kiefer, Michael A. Riley, and Anthony Chemero, "Trading Perception and Action for Complex Cognition: Application of Theoretical Principles from Ecological Psychology to the Design of Interventions for Skill Learning," in *Handbook of Embodied Cognition and Sport Psychology*, ed. Massimiliano Cappuccio (MIT Press, 2019), 47.

9. E. Von Holst, "Die Koordination der Bewegung bei den Arthropoden in Abhängigkeit von zentralen und peripheren Bedingungen," *Biological Reviews*, 10, no. 2 (1935): 234–61. For good summaries in English, see P. N. Kugler and M. T. Turvey. *Information, Natural Law, and the Self-Assembly of Rhythmic Movement* (Routledge, 1987); Janna Iverson and Esther Thelen, "Hand, Mouth and Brain: The Dynamic Emergence of Speech and Gesture," *Journal of Consciousness Studies* 6, no. 11–12 (1999): 19–40.

10. Hermann Haken, J. A. Scott Kelso, and Herbert Bunz, "A Theoretical Model of Phase Transitions in Human Hand Movements," *Biological Cybernetics* 51, no. 5 (1985): 347–56.

11. J. A. Scott Kelso, "Phase Transitions and Critical Behavior in Human Bimanual Coordination," *American Journal of Physiology-Regulatory, Integrative and Comparative Physiology* 246, no. 6 (1984): R1000–R1004.

12. For a more recent review, see J. A. Scott Kelso, "The Haken–Kelso–Bunz (HKB) Model: From Matter to Movement to Mind," *Biological Cybernetics* 115 (2021): 305–22.

## 5. SYNERGIES AND THE INTERTWINED SELF

1. J. A. Scott Kelso, "The Haken–Kelso–Bunz (HKB) Model: From Matter to Movement to Mind," *Biological Cybernetics* 115 (2021): 309.

2. Hermann Haken, "Introduction to Synergetics," in *Synergetics: Cooperative Phenomena in Multi-Component Systems*, ed. Hermann Haken (Vieweg+Teubner, 1973), 9–19; J. A. Scott Kelso, "Synergies: Atoms of Brain and Behavior," in *Progress in Motor Control: A Multidisciplinary Perspective*, ed. Dagmar Sternad (Springer, 2009), 83–91.

3. Michael Riley, M. J. Richardson, K. Shockley, and V. C. Ramenzoni, "Interpersonal Synergies," *Frontiers in Psychology* 2 (2011): 38.

4. Synergies exhibit contextual emergence in which contextual, systemic stability conditions enable and constrain the behavior of system components so that emergent features arise. See Robert Bishop, Michael Silberstein, and Mark Pexton, *Emergence in Context* (Oxford University Press, 2022).

5. Guy Van Orden, John Holden, and Michael T. Turvey, "Self-Organization of Cognitive Performance," *Journal of Experimental Psychology: General* 132 (2003): 331–51; Guy Van Orden, John Holden, and Michael T. Turvey, "Human Cognition and $1/f$ Scaling," *Journal of Experimental Psychology: General* 134 (2005): 117–23; Michael Riley and John Holden, "Dynamics of Cognition," *WIREs Cognitive Science* 3 (2012):593–606.

6. C-K. Peng, Shlomo Havlin, H. Eugene Stanley, and Ary L. Goldberger, "Quantification of Scaling Exponents and Crossover Phenomena in Nonstationary Heartbeat Time Series," *Chaos: An Interdisciplinary Journal of Nonlinear Science* 5, no. 1 (1995): 82–87.

7. C-K. Peng, J. M. Hausdorff, J. E. Mietus, S. Havlin, H. E. Stanley, and A. L. Goldberger, "Fractals in Physiological Control: From Heart Beat to Gait," in *Lévy Flights and Related Topics in Physics: Proceedings of the International Workshop Held at Nice, France, 27–30 June 1994*, ed. Michael F. Shlesinger, George M. Zaslavsky, and Uriel Frisch. (Springer Berlin Heidelberg, 1995), 313–330.

8. Steven L. Bressler and J. A. Scott Kelso, "Cortical Coordination Dynamics and Cognition," *Trends in Cognitive Sciences* 5, no. 1 (2001): 26–36; Walter J. Freeman, "Origin, Structure, and Role of Background EEG Activity. Part 4: Neural Frame Simulation," *Clinical Neurophysiology* 117, no. 3 (2006): 572–89.

9. After reading a draft of this, Fred Cummins points out that there are many ways to describe and model the results that I discuss here. The methods outlined here are, at root, quantitative. For another, Michael Kimmel has developed a qualitative approach to thinking about

synergies. Cummins's own work here is valuable, having both quantitative and qualitative aspects. See Fred Cummins, *The Ground from Which We Speak: Joint Speech and the Collective Subject* (Cambridge Scholars Publishing, 2019); M. Kimmel, "The Micro-Genesis of Interpersonal Synergy," *Ecological Psychology* 33 (2021): 106–45.

10. Van Orden et al., "Self-Organization of Cognitive Performance."
11. Y. Chen, M. Ding, and J. A. Scott Kelso, "Origin of Timing Errors in Human Sensorimotor Coordination," *Journal of Motor Behavior* 33 (1997): 3–8.
12. Joseph W. Buchanan, *Elements of Biology: With Special Reference to Their Role in the Lives of Animals* (Harper & Brothers, 1933). Thanks to Luis Favela for finding these pictures. See also Luis H. Favela and Anthony Chemero, "Ontologically Plural Systems: A Case Study from the Life Sciences," in *Methodology of Situated Cognition Research*, ed. M.-O. Casper and G. Artese (Springer Verlag, 2023).
13. J. Sen and D. McGill, "Fractal Analysis of Heart Rate Variability as a Predictor of Mortality: A Systematic Review and Meta-Analysis," *Chaos: An Interdisciplinary Journal of Nonlinear Science* 28 (2018): 7.
14. Van Orden et al., "Self-Organization of Cognitive Performance"; Van Orden et al., "Human Cognition and 1/*f* Scaling"; John Holden, Guy Van Orden, and Michael T. Turvey, "Dispersion of Response Times Reveals Cognitive Dynamics," *Psychological Review* 116, no. 2 (2009): 318; Riley and Holden, "Dynamics of Cognition."
15. Martin Heidegger, *Being and Time* (HarperCollins 1962); Dobromir G. Dotov, Lin Nie, and Anthony Chemero. "A Demonstration of the Transition from Ready-to-Hand to Unready-to-Hand." *PLoS One* 5, no. 3 (2010): e9433; Dobromir Dotov, Lin Nie, Kevin Wojcik, Anastasia Jinks, Xen Yu, and Anthony Chemero, "Cognitive and Movement Measures Reflect the Transition to Presence-at-Hand," *New Ideas in Psychology* 45 (2017): 1–10; L. Nie, D. Dotov, and A. Chemero, "Readiness-to-Hand, Extended Cognition, and Multifractality," in *Proceedings of the Annual Meeting of the Cognitive Science Society* 33 (2011): 1835–40.
16. Bruce West, *Where Medicine Went Wrong: Rediscovering the Path to Complexity* (World Scientific, 2006).
17. Richard Schmidt, Claudia Carrello, and Michael T. Turvey, "Phase Transitions and Critical Fluctuations in the Visual Coordination of Rhythmic Movements Between People," *Journal of Experimental Psychology: Human Perception and Performance*, 16 (1990): 227.

18.  Michael J. Richardson, Keri Marsh, Robert Isenhower, Justin Goodman, and Richard Schmidt, "Rocking Together: Dynamics of Intentional and Unintentional Interpersonal Coordination," *Human Movement Science* 26 (2007): 867–91.

19.  Patrick Nalepka, Chirstopher Riehm, Carl Mansour, Anthony Chemero, and Michael J. Richardson, "Investigating Strategy Discovery and Coordination in a Novel Virtual Sheep Herding Game Among Dyads," in *Proceedings of the 37th Annual Meeting of the Cognitive Science Society* (2015); Patrick Nalepka, Rachel Kallen, Anthony Chemero, Elliot Saltzman, and Michael J. Richardson, "Herd Those Sheep: Emergent Multiagent Coordination and Behavioral-Mode Switching," *Psychological Science* 28, no. 5 (2017): 630–50; Patrick Nalepka, Maurice Lamb, Rachel Kallen, et al., "Human Social Motor Solutions for Human–Machine Interaction in Dynamical Task Contexts," *Proceedings of the National Academy of Sciences* 116, no. 4 (2019): 1437–46; Patrick Nalepka, Paula Silva, Rachel Kallen, et al., "Task Dynamics Define the Contextual Emergence of Human Corralling Behaviors," *PLOS One* 16, no. 11 (2021): e0260046.

20.  Ashley Walton, Michael J. Richardson, and Anthony Chemero, "Self-Organization and Semiosis in Jazz Improvisation," *International Journal of Signs and Semiotic Systems (IJSSS)* 3, no. 2 (2014): 12–25; Ashely E. Walton, Michael J. Richardson, Peter Langland-Hassan, and Anthony Chemero, "Improvisation and the Self-Organization of Multiple Musical Bodies." *Frontiers in Psychology* 6 (2015): 313; Ashley E. Walton, Auriel Washburn, Peter Langland-Hassan, Anthony Chemero, Heidi Kloos, and Michael J. Richardson, "Creating Time: Social Collaboration in Music Improvisation," *Topics in Cognitive Science* 10, no. 1 (2018): 95–119.

21.  We used several other methods that won't be discussed here. See the appendix and Walton et al., "Creating Time."

22.  Ginevra Castellano, Marcello Mortillaro, Antonio Camurri, Gualtiero Volpe, and Klaus Scherer, "Automated Analysis of Body Movement in Emotionally Expressive Piano Performances," *Music Perception* 26, no. 2 (2008): 103–119; Jay Juchniewicz, "The Influence of Physical Movement on the Perception of Musical Performance." *Psychology of Music* 36, no. 4 (2008): 417–27.

23.  Kevin Shockley, Marie-Vee Santana, and Carol A. Fowler, "Mutual Interpersonal Postural Constraints Are Involved in Cooperative Conversation," *Journal of Experimental Psychology: Human Perception and Performance* 29, no. 2 (2003): 326.

24. Richardson et al., "Rocking Together."
25. Fabian Ramseyer and Wolfgang Tschacher, "Synchrony in Dyadic Psychotherapy Sessions," *Simultaneity: Temporal Structures and Observer Perspectives* (2008): 329–47.
26. Alexandra Paxton and Rick Dale, "Argument Disrupts Interpersonal Synchrony," *Quarterly Journal of Experimental Psychology* 66, no. 11 (2013): 2092–2102.
27. Riccardo Fusaroli, Joanna Rączaszek-Leonardi, and Kristian Tylén. "Dialog as Interpersonal Synergy," *New Ideas in Psychology* 32 (2014): 147–57.
28. Bruce J. West, Elvis L. Geneston, and Paolo Grigolini, "Maximizing Information Exchange Between Complex Networks," *Physics Reports* 468, no. 1–3 (2008): 1–99.
29. Drew H. Abney, Alexandra Paxton, Rick Dale, and Christopher T. Kello, "Complexity Matching in Dyadic Conversation," *Journal of Experimental Psychology: General* 143, no. 6 (2014): 2304.
30. Herbert H. Clark, *Using Language* (Cambridge University Press, 1996).

## 6. SOCIAL ONTOLOGY, REPRESENTATION HUNGER, AND THE INTERTWINED SELF

The discussion in this chapter is focused on the cognitive sciences. Zach Peck and I make a different, more biological case for group cognition. We argue that the default cases of minds and agents are already groups. Human agents, for example, are holobionts; that is, they (we!) are symbiotic, multispecies conglomerates that act as synergies. In his dissertation, Peck argues that this is also the best way to understand human-machine collaborations. Zachary Peck and Anthony Chemero, "Questioning Two Common Assumptions Concerning Group Agency and Group Cognition," in *Proceedings of the Annual Meeting of the Cognitive Science Society* 46 (2024): 5266–72.

1. See, for example, work by Philip Petit, Margaret Gilbert, John Sutton, Deb Tollefsen, Bryce Huebner, Orestis Palermos, Georg Theiner and their coauthors.
2. William McDougal, *The Group Mind* (Putnam, and Sons, 1921), ix.
3. Margaret Gilbert, *On Social Facts* (Princeton University Press, 1992).
4. Michael E. Bratman, "Shared Cooperative Activity," *The Philosophical Review* 101, no. 2 (1992): 327–41.

5.  John Sutton, "Between Individual and Collective Memory: Coordination, Interaction, Distribution," *Social Research: An International Quarterly* 75, no. 1 (2008): 23–48; Georg Theiner, Colin Allen, and Robert L. Goldstone, "Recognizing Group Cognition," *Cognitive Systems Research* 11, no. 4 (2010): 378–95; Deborah Tollefsen and Rick Dale, "Naturalizing Joint Action: A Process-Based Approach," *Philosophical Psychology* 25, no. 3 (2012): 385–407; Deborah P. Tollefsen, Rick Dale, and Alexandra Paxton, "Alignment, Transactive Memory, and Collective Cognitive Systems," *Review of Philosophy and Psychology* 4 (2013): 49–64; Bryce Huebner, *Macrocognition: A Theory of Distributed Minds and Collective Intentionality* (Oxford University Press, 2013); S. Orestis Palermos, "The Dynamics of Group Cognition," *Minds and Machines* 26, no. 4 (2016): 409–40; Robert L. Goldstone and Georg Theiner, "The Multiple, Interacting Levels of Cognitive Systems (MILCS) Perspective on Group Cognition," *Philosophical Psychology* 30, no. 3 (2017): 338–72.

6.  E. Hutchins, *Cognition in the Wild* (MIT Press, 1995).

7.  Theiner et al., "Recognizing Group Cognition."

8.  Andy Clark and David Chalmers, "The Extended Mind," *Analysis* 58, no. 1 (1998): 8.

9.  Daniel M. Wegner, "Transactive Memory: A Contemporary Analysis of the Group Mind," In *Theories of Group Behavior*, ed. Brian Mullen and George R. Goethals (Springer, 1987).

10. There has been some pushback against this by authors like Eric Thompson, Gualtiero Piccinini, and Russ Poldrack. See Eric Thomson, and Gualtiero Piccinini, "Neural Representations Observed," *Minds and Machines* 28 (2018): 191–235; Russell A. Poldrack, "The Physics of Representation," *Synthese* 199, no. 1 (2021): 1307–25. For responses to this pushback, Michael L. Anderson, and Heather Champion, "Some Dilemmas for an Account of Neural Representation: A Reply to Poldrack," *Synthese* 200, no. 2 (2022): 169.

11. Here, I am using the term "group" in the way that participants in the debates use it. Fred Cummins points out that "group" in discussions of group minds or group cognitive systems is vague and undertheorized. See Fred Cummins and Luciana Longo, "The Empirical Discovery of Domains of Assembly and Communion," *Language Sciences* 100 (2023): 101586.

12. K. Ludwig, "Is Distributed Cognition Group Level Cognition?," *Journal of Social Ontology* 1, no. 2 (2015): 189–224, https://doi.org/10.1515/jso

-2015-0001; K. Ludwig, *From Individual to Plural Agency: Collective Action*, vol. 1 (Oxford University Press, 2016); K. Ludwig, *From Plural to Institutional Agency: Collective Action II* (Oxford University Press, 2017).

13. Ludwig, "Is Distributed Cognition Group Level Cognition?," 214–215.

14. Drawing on work in biology, Zachary Peck and I question whether individual rather than group agency is the norm in biology. Peck and Chemero, "Questioning Two Assumptions."

15. Edward Baggs, "All Affordances Are Social: Foundations of a Gibsonan Social Ontology," *Ecological Psychology* 33, no. 3–4 (2021): 257–78.

16. Natalie Sebanz and Günther Knoblich, "Progress in Joint-Action Research," *Current Directions in Psychological Science* 30, no. 2 (2021): 138–43.

17. Tollefsen et al., "Alignment, Transactive Memory, and Collective Cognitive Systems."

18. Andy Clark, "Out of Our Brains," *New York Times*, December 12, 2010 (2010): 12.

19. Fred Cummins, Where Is My Mind? July 1, 2015, *One Second Per Second* (blog), https://onesecondpersecond.wordpress.com/2015/07/01/the-brain -in-the-ass-hypothesis-where-is-my-mind/.

20. Fred Cummins, *The Ground from Which We Speak: Joint Speech and the Collective Subject* (Cambridge Scholars Publishing, 2019).

21. Diego Cosmelli and Evan Thompson, "Embodiment or Envatment? Reflections on the Bodily Basis of Consciousness," in *Enaction: Towards a New Paradigm for Cognitive Science*, ed. John Stewart, Olivier Gapenne, and Ezequiel Di Paolo (MIT Press, 2010).

22. Cummins, *The Ground from which We Speak*, 91.

23. Andy Clark and J. Toribio, "Doing Without Representing?," *Synthese* 101 (1994): 401–31.

24. Damian G. Stephen and James A. Dixon, "Strong Anticipation: Multifractal Cascade Dynamics Modulate Scaling in Synchronization Behaviors," *Chaos, Solitons & Fractals* 44, no. 1–3 (2011): 160–68.

25. Iris van Rooij, Raoul M. Bongers, and F. G. Haselager, "A Non-Representational Approach to Imagined Action," *Cognitive Science* 26, no. 3 (2002): 345–75.

26. Peter Langland-Hassan, *Explaining Imagination* (Oxford University Press, 2020).

27. Margaret A. Boden, *The Creative Mind: Myths and Mechanisms* (1990; repr. Psychology Press, 2003).

28. Emilien Dereclenne, "Enacting Imagination: At the Crossroads of Philosophy of Cognition and Philosophy of Technics" (PhD diss., Université de Technologie de Compiègne, 2022).

29. Tim Ingold, "The Textility of Making," *Cambridge Journal of Economics* 34 (2010): 91–102; Tim Ingold, *Making: Anthropology, Archaeology, Art and Architecture* (Routledge, 2013).

30. Ingold, *Making*, 94.

31. Ingold, "The Textility of Making," 134.

32. Lambros Malafouris, *How Things Shape the Mind* (MIT Press, 2013).

33. Lambros Malafouris, "Thinking as 'Thinging': Psychology with Things," *Current Directions in Psychological Science* 29, no. 1 (2020): 3–8.

34. Malafouris, "Thinking as 'thinging,'" 4.

35. Chris Baber, *Embodying Design: An Applied Science of Radical Embodied Cognition* (MIT Press, 2022); Chris Baber, Anthony Chemero, and Jamie Hall, "What the Jeweller's Hand Tells the Jeweller's Brain: Tool Use, Creativity, and Embodied Cognition," *Philosophy & Technology* 32 (2019): 283–302; Tim Elmo Feiten, Zachary Peck, Kristopher Holland, and Anthony Chemero, "Constructive Constraints: On the Role of Chance and Complexity in Artistic Creativity," *Possibility Studies & Society* 1, no. 3 (2023): 311–23; Wendy Ross, Vlad Glăveanu, and Anthony Chemero, "The Illusion of Freedom," *Constraints in Creativity* 13 (2024): 166; Sune Vork Steffensen and Frédéric Vallée-Tourangeau, "An Ecological Perspective on Insight Problem Solving," *Insight* (2018): 169–90.

36. Rob Withagen and John van der Kamp, "An Ecological Approach to Creativity in Making," *New Ideas in Psychology* 49 (2018): 1–6.

37. Tim Ingold, "Anthropological Affordances," in *Affordances in Everyday Life: A Multidisciplinary Collection of Essays*, ed. Zakariah Djebbara (Springer International, 2022).

38. Vlad P. Glăveanu, "Rewriting the Language of Creativity: The Five A's Framework," *Review of General Psychology* 17, no. 1 (2013): 69–81; Wendy Ross and Frédéric Vallée-Tourangeau, "Microserendipity in the Creative Process," *Journal of Creative Behavior* 55, no. 3 (2021): 661–72.

39. Michael J. Spivey, "Cognitive Science Progresses Toward Interactive Frameworks," *Topics in Cognitive Science* 15, no. 2 (2023): 219–54. See also Mark Dingemanse, Andreas Liesenfeld, Marnix Rasenberg, et al., "Beyond Single-Mindedness: A Figure-Ground Reversal for the Cognitive Sciences," *Cognitive Science* 47, no. 1 (2023): e13230.

## 7. THE PRAGMATIST TRADITION AND INNER SPEECH

Much of this chapter is drawn from a paper I coauthored with Gui Sanches de Oliveira and Vicente Raja. See Guilherme Sanches de Oliveira, Vicente Raja, and Anthony Chemero, "Radical Embodied Cognitive Science and 'Real Cognition,'" *Synthese* 198, no. Suppl 1 (2021): 115–36. Thanks to Gui and Vicente for allowing me to use it here. That paper also includes a good deal of discussion about how to understand the activity of the brain from an ecological-enactive point of view. See Michael Anderson, *After Phrenology* (MIT Press, 2014); Vicente Raja and Michael L. Anderson, "Radical Embodied Cognitive Neuroscience," *Ecological Psychology* 31, no. 3 (2019): 166–81; Vicente Raja, "A Theory of Resonance: Towards an Ecological Cognitive Architecture," *Minds and Machines* 28 (2018): 29–51; Luis H. Favela, *The Ecological Brain: Unifying the Sciences of Brain, Body, and Environment* (Taylor & Francis, 2023).

1. Louis Menand has written an accessible history of this club and pragmatism more generally. See Louis Menand, *The Metaphysical Club: A Story of Ideas in America* (Macmillan, 2002).

2. Matthew Crippen and Jay Schulkin provide excellent coverage of pragmatism and its relation to both phenomenology and embodied cognition. See Matthew Crippen and Jay Schulkin, *Mind Ecologies: Body, Brain, and World* (Columbia University Press, 2020).

3. Henry Cowles explains how John Dewey's observations of children learning became a model for what is now called the scientific method. The next paragraphs on Mead are influenced by Cowles's work and Karen Hanson's *The Self Imagined*. See Henry M. Cowles, *The Scientific Method: An Evolution of Thinking from Darwin to Dewey* (Harvard University Press, 2020); Karen Hanson, *The Self Imagined: Philosophical Reflections on the Social Character of Psyche* (Routledge 1988).

4. John Dewey, *Experience and Nature* (Open Court, 1925), 162.

5. George Herbert Mead, "The Social Settlement: Its Basis and Function," *University of Chicago Record* 12, no. 8 (1907): 108–110.

6. Mead, "The Social Settlement," 110.

7. George Herbert Mead, *Mind, Self, and Society.*(University of Chicago Press, 1934), 133.

8. There is much more to be said about Mead's understanding of mind and self. See Mitchell Aboulifia, "George Herbert Mead and the Unity of the Self," *European Journal of Pragmatism and American Philosophy* 8,

no. 1 (2016): 1–15. For more on the relationship between Mead and enactive-ecological psychology, see Louise Barrett, "Enactivism, Pragmatism . . . Behaviorism?," *Philosophical Studies* 176 (2018): 807–818; Shaun Gallagher, *Action and Interaction* (Oxford University Press, 2020).

9. Lev S. Vygotsky, *Mind in Society: The Development of Higher Psychological Processes*, vol. 86 (Harvard University Press, 1978). See also Harry Daniels, Michael Cole, and James V. Wertsch, eds., *The Cambridge Companion to Vygotsky* (Cambridge University Press, 2007).

10. Dimitris Bolis and Leonard Schilbach also draw on Vygotsky. See Dimitris Bolis and Leonard Schilbach, "'I Interact Therefore I Am': The Self as a Historical Product of Dialectical Attunement," *Topoi* 39 (2020): 521–34.

11. John Dewey, *Impressions of Soviet Russia and the Revolutionary World: Mexico—China—Turkey*, vol. 10 (New Republic, 1929).

12. William James, *The Principles of Psychology* (Henry Holt, 1890).

13. Anne Edwards, "An Interesting Resemblance: Vygotsky, Mead, and American Pragmatism," in *The Cambridge Companion to Vygotsky*, ed. Harry Daniels, Michael Cole, and James V. Wertsch (Cambridge University Press, 2007); Jaan Valsiner and Rene Van der Veer, "On the Social Nature of Human Cognition: An Analysis of the Shared Intellectual Roots of George Herbert Mead and Lev Vygotsky," *Journal for the Theory of Social Behaviour* 18, no. 1 (1988): 117–36.

14. Bas C. Van Fraassen, *The Scientific Image* (Oxford University Press, 1980); Bas C. Van Fraassen, *Scientific Representation: Paradoxes of Perspective* (Oxford University Press, 2008); Michael Weisberg, *Simulation and Similarity: Using Models to Understand the World* (Oxford University Press, 2012); Margaret Morrison, *Reconstructing Reality: Models, Mathematics, and Simulations* (Oxford University Press, 2015).

15. Margaret Morrison and Mary S. Morgan, "Models as Mediating Instruments," in *Models as Mediators: Perspectives on Natural and Social Science*, ed. M. S. Morgan and M. Morrison (Cambridge University Press, 1999); T. Knuuttila, "Modeling and Representing: An Artefactual Approach to Model-Based Representation," *Studies in History and Philosophy of Science* 42, no. 2 (2011): 262–71; Alistair Isaac, "Modeling Without Representation," *Synthese* 190, no. 16 (2013): 3611–23; Guilherme Sanches de Oliveira, "Theory, Practice, and Non-Reductive (Meta) Science," *Australasian Philosophical Review* 2, no. 2 (2018): 199–203; Guilherme Sanches de Oliveira, "Radical Aartifactualism," *European Journal for Philosophy of Science* 12, no. 2 (2022): 36.

16. Michael J. B. Krieger, Jean-Bernard Billeter, and Laurent Keller, "Ant-Like Task Allocation and Recruitment in Cooperative Robots," *Nature* 406, no. 6799 (2000): 992–95; Richard Reeve, Barbara Webb, Andrew Horchler, Giacomo Indiveri, and Roger Quinn, "New Technologies for Testing a Model of Cricket Phonotaxis on an Outdoor Robot," *Robotics and Autonomous Systems* 51, no. 1 (2005): 41–54.

17. Isaac, "Modeling Without Representation."

## 8. REORIENTING ETHICS AND POLITICAL THEORY AROUND THE INTERTWINED SELF

1. Francisco J. Varela, *Ethical Know-How: Action, Wisdom, and Cognition* (Stanford University Press, 1999).

2. Gilles Deleuze, *Difference and Repetition*, trans. Paul Patton (Columbia University Press, 1994; Gilles Deleuze, *Spinoza: Practical Philosophy* trans. Robert Hurley (City Lights, 1988); Gilles Deleuze and Félix Guattari, *A Thousand Plateaus*, tran. Brian Massumi (Bloomsbury Academic, 2013).

3. Moira Gatens, "Feminism as 'Password': Re-Thinking the 'Possible' with Spinoza and Deleuze," *Hypatia* 15, no. 2 (2000): 59–75; Daniel W. Smith, *Essays on Deleuze* (Edinburgh University Press, 2012); John Protevi, *Life, War, Earth: Deleuze and the Sciences* (University of Minnesota Press, 2013); Adrian Parr, *Deleuze Dictionary Revised Edition* (Edinburgh University Press, 2010).

4. Protevi, *Life, War, Earth*.

5. Donna J. Haraway, *When Species Meet* (University of Minnesota Press, 2008), 30. Unlike Haraway, Moira Gatens does see feminist possibilities in Deleuze's work. In describing Deleuze's work, she writes: "A micropolitical feminism is able to imagine alternative possible forms of sociability. This power of imagining things otherwise, in concert with the imaginings of compatible others, has the creative power to decompose and recompose the social field, bit by bit, molecule by molecule." Gatens, "Feminism as 'Password,'" 72.

6. Donna J. Haraway, "A Manifesto for Cyborgs," *Socialist Review* 80 (1985): 65–108.

7. Donna J. Haraway, *Staying with the Trouble: Making Kin in the Chthulucene* (Duke University Press, 2016), 104–105.

8. See also Mads J. Dengsø and Michael D. Kirchhoff, "Beyond Individual-Centred 4E Cognition: Systems Biology and Sympoiesis," *Constructivist Foundations* 18, no. 3 (2023): 351–64.

9. Haraway, *Staying with the Trouble*, 58.

10. Moria Gatens, *Feminism and Philosophy: Perspectives on Difference and Equality* (Indiana University Press, 1991).

11. Gatens, *Feminism and Philosophy*, 5.

12. John A. Rawls, *A Theory of Justice* (Harvard University Press, 1971).

13. Lorraine Code, *What Can She Know?: Feminist Theory and the Construction of Knowledge* (Cornell University Press, 1991), 72–73.

14. Annette Baier, *Postures of the Mind: Essays on Mind and Morals* (University of Minnesota Press, 1985).

15. Code, *What Can She Know?*, 72.

16. Code, *What Can She Know?*, 85.

17. Code, *What Can She Know?*, 85.

18. Mason Cash, "Extended Cognition, Personal Responsibility, and Relational Autonomy," *Phenomenology and the Cognitive Sciences* 9 (2010): 645–71; Mason Cash "Cognition Without Borders: 'Third Wave' Socially Distributed Cognition and Relational Autonomy," *Cognitive Systems Research* 25 (2013): 61–71; Elena Cuffari, "Habits of Transformation," *Hypatia* 26, no. 3 (2011): 535–53; Michelle Maiese, "Essentially Embodied, Emotive, Enactive Social Cognition," in *Embodiment, Emotion, and Cognition*, (Palgrave Macmillan, 2011), 151–84; Michelle Maiese, "Embodiment, Sociality, and the Life Shaping Thesis," *Phenomenology and the Cognitive Sciences* 18, no. 2 (2019): 353–74; Hanne De Jaegher, "Rigid and Fluid Interactions with Institutions," *Cognitive Systems Research* 25 (2013): 19–25; Sara Heinämaa, "Sex, Gender and Embodiment," in *The Oxford Handbook of Contemporary Phenomenology*, ed. Dan Zahavi (Oxford University Press, 2012); Nick Brancazio, "Gender and the Senses of Agency," *Phenomenology and the Cognitive Sciences* (2018), https://doi.org/10.1007/s11097-018-9581-z; Nick Brancazio, "Being Perceived and Being 'Seen': Interpersonal Affordances, Agency, and Selfhood," *Frontiers in Psychology* 11 (2020): 1750; Laura Candiotto and Hanne De Jaegher, "Love In-Between," *Journal of Ethics* (2021), https://doi.org/10.1007/s10892-020-09357-9; Anya Daly, *Merleau-Ponty and the Ethics of Intersubjectivity* (Palgrave Macmillan, 2016); Anya Daly. "The Declaration of Interdependence! Feminism, Grounding and Enactivism," *Human Studies* (2021), https://doi.org/10.1007/s10746-020-09570-3; Janna van Grunsven, "Enactivism, Second-Person Engagement and Personal Responsibility," *Phenomenology and the Cognitive Sciences* 17 (2018): 131–56; Shen-Yi Liao and Bryce Huebner, "Oppressive Things," *Philosophy*

*and Phenomenological Research* 103, no. 1 (2021): 92–113; Shen-Yi Liao and Vanessa Carbonell, "Materialized Oppression in Medical Tools and Technologies," *American Journal of Bioethics: AJOB* (2022): 1–15; Juan M. Loaiza, "From Enactive Concern to Care in Social Life: Towards an Enactive Anthropology of Caring," *Adaptive Behavior* 27, no. 1 (2019): 17–30; Geoffrey Dierckxsens, ed., *Ethical Dimensions of Enactive Cognition— Perspectives on Enactivism, Bioethics and Applied Ethics*, special issue of *Topoi* 41, no. 2 (2022): 235–437.

19. Catriona Mackenzie and Natalie Stoljar, eds., *Relational Autonomy: Feminist Perspectives on Autonomy, Agency, and the Social Self* (Oxford University Press, 2000), 3.

20. Lauren Freeman makes a case for relational autonomy based on Heidegger's *Being and Time*, which I discuss in chapter 1. See Lauren Freeman, "Reconsidering Relational Autonomy: A Feminist Approach to Selfhood and the Other in the Thinking of Martin Heidegger," *Inquiry* 54, no. 4 (2011): 361–83.

21. Cash, "Extended Cognition"; Cash, "Cognition Without Borders."

22. Cash, "Extended Cognition."

23. Cash, "Extended Cognition."

24. Michelle Maiese, *Autonomy, Enactivism, and Mental Disorder: A Philosophical Account* (Taylor & Francis, 2022).

25. See, for example, Julia Annas, *Intelligent Virtue* (Oxford University Press, 2011).

26. Erand Jayawickreme and Anthony Chemero, "Ecological Moral Realism: An Alternative Theoretical Framework for Studying Moral Psychology," *Review of General Psychology* 12, no. 2 (2008): 118–26.

27. Carol Gilligan, "In a Different Voice: Women's Conceptions of Self and of Morality," *Harvard Educational Review* 47, no. 4 (1977): 481–517.

28. Carol Gilligan, *In a Different Voice: Psychological Theory and Women's Development* (Harvard University Press, 1982).

29. Robin S. Dillon, "Feminist Virtue Ethics," in *The Routledge Companion to Feminist Philosophy*, ed. Ann Garry, Serene J. Khader, and Alison Stone (Routledge, 2017).

30. Joan C. Tronto, "Beyond Gender Difference to a Theory of Care," *Signs: Journal of Women in Culture and Society* 12, no. 4 (1987): 644–63.

31. Anjali Dutt and Danielle Kohfeldt, "Towards a Liberatory Ethics of Care Framework for Organizing Social Change," *Journal of Social and Political Psychology* 6, no. 2 (2018): 575–90.

32. M. Órnelas, "Moral Affordances" (PhD diss., University of Cincinnati, 2024).
33. van Grunsven, "Enactivism, Second-Person Engagement"; Loaiza, "From Enactive Concern to Care in Social Life."
34. Paul Bloom, *Against Empathy: The Case for Rational Compassion* (Random House, 2016).
35. Karl Marx, *Theses on Feuerbach*, trans. C. Smith (Marx-Engels Internet Archive 2002).

## 9. CODA: BLANKS AMONG US

A version of this chapter was published in Anthony Chemero, "LLMs Differ from Human Cognition Because They Are Not Embodied," *Nature Human Behaviour 7*, no. 11 (2023): 1828–29.

1. Abeba Birhane, "The Unseen Black Faces of AI Algorithms," *Nature* 601 (2022): 451–52; Abeba Birhane and Marek McGann, "Large Models of What? Mistaking Engineering Achievements for Human Linguistic Agency," *Language Sciences* 106 (2024): 101672.
2. J. Haugeland, *Having Thought: Essays in the Metaphysics of Mind* (Harvard University Press, 1998).

## APPENDIX FOR PEOPLE WHO LIKE MATH

1. Per Bak, Chao Tang, and Kurt Wiesenfeld, "Self-Organized Criticality: An Explanation of the $1/f$ Noise," *Physical review letters* 59, no. 4 (1987): 381.
2. Michael Riley, Michael J. Richardson, Kevin Shockley, and Vanessa C. Ramenzoni, "Interpersonal Synergies," *Frontiers in Psychology* 2 (2011).
3. M. T. Turvey, H. L. Fitch, and B. Tuller, "The Bernstein perspective: I. The Problems of Degrees of Freedom and Context-Conditioned Variability," *Human Motor Behavior: An Introduction*, ed. J. A. S Kelso, (Erlbaum, 1982), 239–52; Mark L. Latash, *Synergy* (Oxford University Press, 2008).
4. Andreas Daffertshofer, C. J. Lamoth, O. G. Meijer, and P. J. Beek, "PCA in Studying Coordination and Variability: A Tutorial," *Clinical Biomechanics* 19, no. 4 (2004): 415–28.
5. R. C. Schmidt, Lin Nie, Alison Franco, and Michael J. Richardson, "Bodily Synchronization Underlying Joke Telling," *Frontiers in Human Neuroscience* 8 (2014): 633; Auriel Washburn, Mariana DeMarco, Simon

de Vries, et al., "Dancers Entrain More Effectively Than Non-Dancers to Another Actor's Movements," *Frontiers in Human Neuroscience* 8 (2014): 800.

6. Johann Issartel, Ludovic Marin, Philippe Gaillot, Thomas Bardainne, and Marielle Cadopi, "A Practical Guide to Time—Frequency Analysis in the Study of Human Motor Behavior: The Contribution of Wavelet Transform," *Journal of Motor Behavior* 38, no. 2 (2006): 139–59.

7. This analysis outputs extraordinarily informative images concerning the details of the coordination of the players. See Ashley E. Walton, Michael J. Richardson, Peter Langland-Hassan, and Anthony Chemero, "Improvisation and the Self-Organization of Multiple Musical Bodies," *Frontiers in Psychology* 6 (2015).

8. Ginevra Castellano, Marco Mortillaro, Antonio Camurri, Gualtiero Volpe, and Klaus Scherer, "Automated Analysis of Body Movement in Emotionally Expressive Piano Performances," *Music Perception* 26, no. 2 (2008): 103–119, https://doi.org/10.1525/mp.2008.26.2.103. Jay Juchniewicz, "The Influence of Physical Movement on the Perception of Musical Performance," *Psychology of Music* 36, no. 4 (2008): 417–27, https://doi.org/10.1177/0305735607086046.

9. C. T. Kello, S. D. Bella, B. Médé, and R. Balasubramaniam, "Hierarchical Temporal Structure in Music, Speech and Animal Vocalizations: Jazz Is Like a Conversation, Humpbacks Sing Like Hermit Thrushes," *Journal of the Royal Society Interface* 14, no. 135 (2017): 20170231.

**META-APPENDIX: THE CONTROVERSY OVER 1/F NOISE**

A version of this appendix was originally published in *Topics in Cognitive Science*. See Anthony Chemero, "Abduction and Deduction in Dynamical Cognitive Science," *Topics in Cognitive Science* (2023).

1. Per Bak, Chao Tang, and Kurt Wiesenfeld, "Self-Organized Criticality: An Explanation of the 1/*f* Noise," *Physical Review Letters* 59, no. 4 (1987): 381.

2. Jerry A. Fodor, *The Modularity of Mind* (MIT Press, 1983); Jerry A. Fodor, *The Mind Doesn't Work That Way: The Scope and Limits of Computational Psychology* (MIT Press, 2001); John Haugeland, *Having Thought: Essays in the Metaphysics of Mind* (Harvard University Press, 1998); Michael T. Turvey, *Lectures on Perception: An Ecological Perspective* (Routledge, 2018).

3. Harry G. Frankfurt, "Peirce's Notion of Abduction," *The Journal of Philosophy* 55, no. 14 (1958): 593–97.

4.  Carl G. Hempel and Paul Oppenheim, "Studies in the Logic of Explanation," *Philosophy of Science* 15, no. 2 (1948): 135–75.
5.  Guy C. van Orden, John G. Holden, and Michael T. Turvey, "Self-Organization of Cognitive Performance," *Journal of Experimental Psychology: General* 132 (2003): 331–51.
6.  Eric-Jan Wagenmakers, Simon Farrell, and Roger Ratcliff, "Human Cognition and a Pile of Sand: A Discussion on Serial Correlations and Self-Organized Criticality," *Journal of Experimental Psychology: General* 134, no. 1 (2005): 108.
7.  Guy Van Orden, John Holden, and Michael T. Turvey, "Human Cognition and 1/f Scaling," *Journal of Experimental Psychology: General* 134 (2005): 117–23.
8.  E. A. Ihlen and B. Vereijken, "Interaction-Dominant Dynamics in Human Cognition: Beyond 1/fα Fluctuation," *Journal of Experimental Psychology: General* 139, no. 3 (2010): 436.
9.  Bruce J. West, Elvis L. Geneston, and Paolo Grigolini, "Maximizing Information Exchange Between Complex Networks," *Physics Reports* 468 (2008): 1–99.
10. Damian G. Stephen and James A. Dixon, "Strong Anticipation: Multifractal Cascade Dynamics Modulate Scaling in Synchronization Behaviors," *Chaos, Solitons & Fractals* 44, no. 1–3 (2011): 160–68; Didier Delignières, Z. M. Almurad, C. Roume, and V. Marmelat, "Multifractal Signatures of Complexity Matching," *Experimental Brain Research* 234 (2016): 2773–85.
11. Lillian M. Rigoli, Daniel Holman, Michael J. Spivey, and Christopher T. Kello, "Spectral Convergence in Tapping and Physiological Fluctuations: Coupling and Independence of 1/f Noise in the Central and Autonomic Nervous Systems," *Frontiers in Human Neuroscience* 8 (2014): 713.
12. Vicente Raja, "A Theory of Resonance: Towards an Ecological Cognitive Architecture," *Minds and Machines* 28 (2018): 29–51; Luis H. Favela, *The Ecological Brain: Unifying the Sciences of Brain, Body, and Environment* (Taylor & Francis, 2023).
13. Alexandra Paxton and Rick Dale, "Argument Disrupts Interpersonal Synchrony," *Quarterly Journal of Experimental Psychology* 66, no. 11 (2013): 2092–2102; Drew H. Abney, Alexandra Paxton, Rick Dale, and Christopher T. Kello, "Complexity Matching in Dyadic Conversation," *Journal of Experimental Psychology: General* 143, no. 6 (2014): 2304–15.

14. Jay Friedenberg, Gordon Silverman, and Michael J. Spivey, *Cognitive Science: An Introduction to the Study of Mind* (Sage, 2021).

15. Rick Dale, "The Possibility of a Pluralist Cognitive Science," *Journal of Experimental and Theoretical Artificial Intelligence* 20, no. 3 (2008): 155–79; Luis H. Favela, Luis H., and Anthony Chemero, "Ontologically Plural Systems: A Case Study from the Life Sciences," in *Methodology of Situated Cognition Research*, ed. Mario-Otto Casper and Giovanni Artese (Springer Verlag, 2023).

# BIBLIOGRAPHY

Abney, Drew H., Alexandra Paxton, Rick Dale, and Christopher T. Kello. "Complexity Matching in Dyadic Conversation." *Journal of Experimental Psychology: General* 143, no. 6 (2014): 2304–15.

Abney, Drew H., Alexandra Paxton, Rick Dale, and Christopher T. Kello. "Cooperation in Sound and Motion: Complexity Matching in Collaborative Interaction." *Journal of Experimental Psychology: General* 150, no. 9 (2021): 1760.

Aboulafia, Mitchell. "George Herbert Mead and the Unity of the Self." *European Journal of Pragmatism and American Philosophy* 8, no. 8-1 (2016): 1–16.

Anderson, Michael L. *After Phrenology.* MIT Press, 2014.

Anderson, Michael L., and Heather Champion. "Some Dilemmas for an Account of Neural Representation: A Reply to Poldrack." *Synthese* 200, no. 2 (2022): 169.

Annas, Julia. *Intelligent Virtue.* Oxford University Press, 2011.

Avramides, Anita. *Other Minds.* Routledge, 2000.

Baber, Chris. *Embodying Design: An Applied Science of Radical Embodied Cognition.* MIT Press, 2022.

Baber, Chris, Anthony Chemero, and Jamie Hall. "What the Jeweller's Hand Tells the Jeweller's Brain: Tool Use, Creativity and Embodied Cognition." *Philosophy & Technology* 32 (2019): 283–302.

Baggs, Edward. "All Affordances Are Social: Foundations of a Gibsonian Social Ontology." *Ecological Psychology* 33, no. 3–4 (2021): 257–78.

Baggs, Edward, and Anthony Chemero. "Radical Embodiment in Two Directions." *Synthese* 198, suppl. 9 (2021): 2175–90.

Baier, Annette. *Postures of the Mind: Essays on Mind and Morals.* University of Minnesota Press, 1985.

Bak, Per, Chao Tang, and Kurt Wiesenfeld. "Self-Organized Criticality: An Explanation of the $1/f$ Noise." *Physical Review Letters* 59, no. 4 (1987): 381.

Barrett, Louise. *Beyond the Brain: How Body and Environment Shape Animal and Human Minds.* Princeton University Press, 2011.

Barrett, Louise. "Enactivism, Pragmatism . . . Behaviorism?" *Philosophical Studies* 176, no. 3 (2018): 807–18.

Barrett, Louise. "Why Behaviorism Isn't Satanism." In *The Oxford Handbook of Comparative Evolutionary Psychology,* ed. Todd K. Shackelford and Jennifer Vonk. Oxford University Press, 2012.

Beer, Randall D. "A Dynamical Systems Perspective on Agent-Environment Interaction." *Artificial Intelligence* 72, no. 1–2 (1995): 173–215.

Birhane, Abeba. "Descartes Was Wrong: A Person Is a Person Through Other Persons." *Aeon,* April 7, 2017.

Birhane, Abeba. "The Unseen Black Faces of AI Algorithms." *Nature* 601 (2022): 451–52.

Bishop, Robert, Michael Silberstein, and Mark Pexton. *Emergence in Context.* Oxford University Press, 2022.

Bloom, Paul. *Against Empathy: The Case for Rational Compassion.* Random House, 2016.

Boden, Margaret A. *The Creative Mind: Myths and Mechanisms.* 2nd ed. Psychology Press, 2003.

Bolis, Dimitris, and Leonhard Schilbach. "'I Interact Therefore I Am': The Self as a Historical Product of Dialectical Attunement." *Topoi* 39 (2020): 521–34.

Boring, Edwin G. *A History of Experimental Psychology.* Appleton-Century, 1929.

Brancazio, Nick. "Being Perceived and Being 'Seen': Interpersonal Affordances, Agency, and Selfhood." *Frontiers in Psychology* 11 (2020): 1750.

Brancazio, Nick. "Gender and the Senses of Agency." *Phenomenology and the Cognitive Sciences* (2018). https://doi.org/10.1007/s11097-018-9581-z.

Brancazio, Nick, and Miguel Segundo-Ortin. "Distal Engagement: Intentions in Perception." *Consciousness and Cognition* 79 (2020): 102897.

Bratman, Michael E. "Shared Cooperative Activity." *Philosophical Review* 101, no. 2 (1992): 327–41.

Bressler, Steven L., and J. A. Scott Kelso. "Cortical Coordination Dynamics and Cognition." *Trends in Cognitive Sciences* 5, no. 1 (2001): 26–36.

Bruineberg, Jelle, Anthony Chemero, and Erik Rietveld. "General Ecological Information Supports Engagement with Affordances for 'Higher' Cognition." *Synthese* 196 (2019): 5231–51.

Bruineberg, Jelle, Rob Withagen, and Ludger van Dijk. "Productive Pluralism: The Coming of Age of Ecological Psychology." *Psychological Review* (2023). https://doi.org/10.1037/rev0000438.

Buchanan, James W. *Elements of Biology: With Special Reference to Their Role in the Lives of Animals*. Harper & Brothers, 1933.

Candiotto, Laura, and Hanne De Jaegher. "Love In-Between." *Journal of Ethics* (2021). https://doi.org/10.1007/s10892-020-09357-9.

Cash, Mason. "Cognition Without Borders: 'Third Wave' Socially Distributed Cognition and Relational Autonomy." *Cognitive Systems Research* 25 (2013): 61–71.

Cash, Mason. "Extended Cognition, Personal Responsibility, and Relational Autonomy." *Phenomenology and the Cognitive Sciences* 9 (2010): 645–71.

Castellano, Ginevra, Marco Mortillaro, Antonio Camurri, Gualtiero Volpe, and Klaus Scherer. "Automated Analysis of Body Movement in Emotionally Expressive Piano Performances." *Music Perception* 26, no. 2 (2008): 103–19. https://doi.org/10.1525/mp.2008.26.2.103.

Chemero, Anthony. "Abduction and Deduction in Dynamical Cognitive Science." *Topics in Cognitive Science*. https://doi.org/10.1111/tops.12692.

Chemero, Anthony. "LLMs Differ from Human Cognition Because They Are Not Embodied." *Nature Human Behaviour* (2023). https://rdcu.be/drzoz.

Chemero, Anthony. *Radical Embodied Cognitive Science*. MIT Press, 2009.

Chemero, Anthony. "Sensorimotor Empathy." *Journal of Consciousness Studies* 23, nos. 5–6 (2016): 138–52.

Chen, Y., M. Ding, and J. A. Scott Kelso. "Origin of Timing Errors in Human Sensorimotor Coordination." *Journal of Motor Behavior* 33 (2001): 3–8.

Chomsky, Noam. *What Kind of Creatures Are We?* Columbia University Press, 2015.

Churchland, Paul M. "Eliminative Materialism and the Propositional Attitudes." *Journal of Philosophy* 78, no. 2 (1981): 67–90.

Clark, Andy. *Being There: Putting Brain, Body, and World Together Again*. MIT Press, 1997.

Clark, Andy. *Mindware: An Introduction to the Philosophy of Cognitive Science.* Oxford University Press, 2000.

Clark, Andy, and David Chalmers. "The Extended Mind." *Analysis* 58, no. 1 (1998): 7–19.

Clark, Andy, and Josefa Toribio. "Doing Without Representing?" *Synthese* 101 (1994): 401–31.

Clark, Herbert H. *Using Language.* Cambridge University Press, 1996.

Code, Lorraine. *What Can She Know? Feminist Theory and the Construction of Knowledge.* Cornell University Press, 1991.

Coey, Charles A., Rachel W. Kallen, Anthony Chemero, and Michael J. Richardson. "Exploring Complexity Matching and Asynchrony Dynamics in Synchronized and Syncopated Task Performances." *Human Movement Science* 62 (2018): 81–104.

Colombetti, Giovanna. *The Feeling Body: Affective Science Meets the Enactive Mind.* MIT Press, 2014.

Cosmelli, Diego, and Evan Thompson. "Embodiment or Envatment? Reflections on the Bodily Basis of Consciousness." In *Enaction: Towards a New Paradigm for Cognitive Science,* ed. John Stewart, Olivier Gapenne, and Ezequiel Di Paolo. MIT Press, 2010.

Cowles, Henry M. *The Scientific Method: An Evolution of Thinking from Darwin to Dewey.* Harvard University Press, 2020.

Crippen, Matthew, and Jay Schulkin. *Mind Ecologies: Body, Brain, and World.* Columbia University Press, 2020.

Cuffari, Elena. "Habits of Transformation." *Hypatia* 26, no. 3 (2011): 535–53.

Cummins, Fred. *The Ground from Which We Speak: Joint Speech and the Collective Subject.* Cambridge Scholars Publishing, 2019.

Daffertshofer, Andreas, Claudine J. C. Lamoth, Onno G. Meijer, and Peter J. Beek. "PCA in Studying Coordination and Variability: A Tutorial." *Clinical Biomechanics* 19, no. 4 (2004): 415–28.

Dale, Rick. "The Possibility of a Pluralist Cognitive Science." *Journal of Experimental and Theoretical Artificial Intelligence* 20, no. 3 (2008): 155–79.

Daly, Anya. "The Declaration of Interdependence! Feminism, Grounding and Enactivism." *Human Studies* (2021). https://doi.org/10.1007/s10746-020-09570-3.

Daly, Anya. *Merleau-Ponty and the Ethics of Intersubjectivity.* Palgrave Macmillan, 2016.

Daniels, Harry, Michael Cole, and James V. Wertsch, eds. *The Cambridge Companion to Vygotsky.* Cambridge University Press, 2007.

de Haan, Sanneke. "The Minimal Self Is a Social Self." In *The Embodied Self: Dimensions, Coherence, and Disorders*, ed. Thomas Fuchs, Heribert Sattel, and Peter Henningsen. Schattauer, 2010.

de Haan, Sanneke, and Leon de Bruin. "Reconstructing the Minimal Self, or How to Make Sense of Agency and Ownership." *Phenomenology and the Cognitive Sciences* 9 (2010): 373–96.

De Jaegher, Hanne. "Rigid and Fluid Interactions with Institutions." *Cognitive Systems Research* 25 (2013): 19–25.

De Jaegher, Hanne, and Ezequiel Di Paolo. "Participatory Sense-Making: An Enactive Approach to Social Cognition." *Phenomenology and the Cognitive Sciences* 6 (2007): 485–507.

De Jaegher, Hanne, Ezequiel Di Paolo, and Shaun Gallagher. "Can Social Interaction Constitute Social Cognition?" *Trends in Cognitive Sciences* 14, no. 10 (2010): 441–47.

Deleuze, Gilles. *Difference and Repetition*. Trans. Paul Patton. Columbia University Press, 1994. Originally published as *Différence et répétition* (Presses Universitaires de France, 1968).

Deleuze, Gilles. *Spinoza: Practical Philosophy*. Trans. Robert Hurley. City Lights Books, 1988. Originally published as *Spinoza: Philosophie pratique* (Éditions de Minuit, 1968).

Deleuze, Gilles, and Félix Guattari. *A Thousand Plateaus: Capitalism and Schizophrenia*. Trans. Brian Massumi. Bloomsbury Academic, 2013. Originally published as *Mille plateaux* (Éditions de Minuit, 1980).

Delignières, Didier, Zeinab M. Almurad, Claire Roume, and Virginie Marmelat. "Multifractal Signatures of Complexity Matching." *Experimental Brain Research* 234 (2016): 2773–85.

Dengsø, Martin J., and Michael D. Kirchhoff. "Beyond Individual-Centred 4E Cognition: Systems Biology and Sympoiesis." *Constructivist Foundations* 18, no. 3 (2023): 351–64. https://constructivist.info/18/3/351.

Dennett, Daniel C. *Consciousness Explained*. Little, Brown, 1991.

Dennett, Daniel C. *The Intentional Stance*. MIT Press, 1987.

Dereclenne, Emmanuel. "Enacting Imagination: At the Crossroads of Philosophy of Cognition and Philosophy of Technics." PhD diss., Université de Technologie de Compiègne, 2022.

Descartes, René. *Discourse on the Method of Rightly Conducting One's Reason and Seeking Truth in the Sciences*. Trans. John Veitch. Project Gutenberg, 2001. Originally published in 1637. https://www.gutenberg.org/files/59/59-h /59-h.htm.

Dewey, John. *Experience and Nature*. Open Court, 1925.

Dewey, John. *Impressions of Soviet Russia and the Revolutionary World: Mexico—China—Turkey*. Vol. 10. New Republic, 1929.

Dierckxsens, Geoffrey, ed. *Ethical Dimensions of Enactive Cognition—Perspectives on Enactivism, Bioethics and Applied Ethics*. Special issue of *Topoi* 41, no. 2 (2022): 235–437.

Dillon, Robin S. "Feminist Virtue Ethics." In *The Routledge Companion to Feminist Philosophy*, ed. Ann Garry, Serene J. Khader, and Alison Stone. Routledge, 2017.

Dingemanse, Mark, Andreas Liesenfeld, Marlou Rasenberg, et al. "Beyond Single-Mindedness: A Figure-Ground Reversal for the Cognitive Sciences." *Cognitive Science* 47, no. 1 (2023): e13230.

Di Paolo, Ezequiel, Thomas Buhrmann, and Xabier Barandiaran. *Sensorimotor Life: An Enactive Proposal*. Oxford University Press, 2017.

Di Paolo, Ezequiel A., Manuel Heras-Escribano, Anthony Chemero, and Marek McGann, eds. *Enaction and Ecological Psychology: Convergences and Complementarities*. Frontiers Media, 2021.

Dotov, Dobromir G., Lin Nie, and Anthony Chemero. "A Demonstration of the Transition from Ready-to-Hand to Unready-to-Hand." *PLoS ONE* 5, no. 3 (2010): e9433.

Dotov, Dobromir, Lin Nie, Kevin Wojcik, Anastasia Jinks, Xiyu Yu, and Anthony Chemero. "Cognitive and Movement Measures Reflect the Transition to Presence-at-Hand." *New Ideas in Psychology* 45 (2017): 1–10.

Dutt, Anjali, and Dina Kohfeldt. "Towards a Liberatory Ethics of Care Framework for Organizing Social Change." *Journal of Social and Political Psychology* 6, no. 2 (2018): 575–90.

Edwards, Anne. "An Interesting Resemblance: Vygotsky, Mead, and American Pragmatism." In *The Cambridge Companion to Vygotsky*, ed. Harry Daniels, Michael Cole, and James V. Wertsch. Cambridge University Press, 2007.

Egbert, Matthew D., Xabier E. Barandiaran, and Ezequiel A. Di Paolo. "Behavioral Metabolution: Metabolism-Based Behavior Enables New Forms of Adaptation and Evolution." In *Artificial Life XII: Proceedings of the Twelfth International Conference on the Synthesis and Simulation of Living Systems*, ed. Harold Fellermann, Mark D'orr, Martin Hanczyc, Lone Ladegaard Laursen, Sarah Maurer, Daniel Merkle, Pierre-Alain Monnard, Kasper Støy, and Steen Rasmussen. MIT Press, 2010.

Evarts, Edward V. "Sherrington's Concept of Proprioception." *Trends in Neurosciences* 4 (1981): 44–46.

Fanon, Frantz. *Black Skin, White Masks*. Trans. Charles Lam Markmann. Grove Press, 1967. Originally published as *Peau noire, masques blancs* (Éditions du Seuil, 1952).

Favela, Luis H. *The Ecological Brain: Unifying the Sciences of Brain, Body, and Environment*. Taylor & Francis, 2023).

Favela, Luis H., Mary J. Amon, Lorena Lobo, and Anthony Chemero. "Empirical Evidence for Extended Cognitive Systems." *Cognitive Science* 45, no. 11 (2021): e13060.

Favela, Luis H., and Anthony Chemero. "Ontologically Plural Systems: A Case Study from the Life Sciences." In *Methodology of Situated Cognition Research*, ed. Mario-Otto Casper and Giovanni Artese. Springer Verlag, 2023.

Fechner, Gustav. *Elements of Psychophysics*. Vol. 1. Trans. Helmut E. Adler. Holt, Rinehart and Winston, 1966. Originally published as *Elemente der Psychophysik* (Breitkopf & Härtel, 1860).

Feiten, Elmo, Zachary Peck, Kristopher Holland, and Anthony Chemero. "Creative Constraints." *Possibility Studies and Society* 1, no. 3 (2023): 311–23.

Fodor, Jerry A. "Methodological Solipsism Considered as a Research Strategy in Cognitive Psychology." *Behavioral and Brain Sciences* 3, no. 1 (1980): 63–73.

Fodor, Jerry A. *The Mind Doesn't Work That Way: The Scope and Limits of Computational Psychology*. MIT Press, 2001.

Fodor, Jerry A. *The Modularity of Mind: An Essay on Faculty Psychology*. MIT Press, 1983.

Fodor, Jerry A. *Psychosemantics: The Problem of Meaning in the Philosophy of Mind*. MIT Press, 1987.

Frankfurt, Harry G. "Peirce's Notion of Abduction." *Journal of Philosophy* 55, no. 14 (1958): 593–97.

Freeman, Walter J. "Origin, Structure, and Role of Background EEG Activity. Part 4: Neural Frame Simulation." *Clinical Neurophysiology* 117, no. 3 (2006): 572–89.

Friedenberg, Jay, Gordon Silverman, and Michael J. Spivey. *Cognitive Science: An Introduction to the Study of Mind*. Sage, 2021.

Froese, Tom, and Ezequiel A. Di Paolo. "Modelling Social Interaction as Perceptual Crossing: An Investigation into the Dynamics of the Interaction Process." *Connection Science* 22, no. 1 (2010): 43–68.

Fuchs, Thomas. *Ecology of the Brain: The Phenomenology and Biology of the Embodied Mind*. Oxford University Press, 2018.

Fuchs, Thomas, and Hanne De Jaegher. "Enactive Intersubjectivity: Partici-patory Sense-Making and Mutual Incorporation." *Phenomenology and the Cognitive Sciences* 8 (2009): 465–86.

Fusaroli, Riccardo, Joanna Rączaszek-Leonardi, and Kristian Tylén. "Dialog as Interpersonal Synergy." *New Ideas in Psychology* 32 (2014): 147–57.

Gabbey, Alan. "Reflections on the Other Minds Problem: Descartes and Oth-ers." In *Sceptics, Millenarians and Jews*, ed. David S. Katz and Jonathan I. Israel. Brill, 1990.

Gallagher, Shaun. *Action and Interaction*. Oxford University Press, 2020.

Gallagher, Shaun. *Enactivist Interventions: Rethinking the Mind*. Oxford Uni-versity Press, 2017.

Gallagher, Shaun. *How the Body Shapes the Mind*. Clarendon Press, 2005.

Gallagher, Shaun. "A Pattern Theory of Self." *Frontiers in Human Neuroscience* 7 (2013): 443.

Gallagher, Shaun. "Philosophical Conceptions of the Self: Implications for Cognitive Science." *Trends in Cognitive Sciences* 4, no. 1 (2000): 14–21.

Gallagher, Shaun. *The Self and Its Disorders* Oxford University Press, 2024.

Gallagher, Shaun. "A Well-Trodden Path: From Phenomenology to Enactiv-ism." *Filosofisk Supplement* 3 (2018): 42–47.

Gallagher, Shaun, and Daniel D. Hutto. "Understanding Others Through Pri-mary Interaction and Narrative Practice." In *The Shared Mind: Perspectives on Intersubjectivity*, ed. Jordan Zlatev, Timothy P. Racine, Chris Sinha, and Esa Itkonen. John Benjamins, 2007.

Gallagher, Shaun, and Tailer G. Ransom. "Artifacting Minds: Material Engage-ment Theory and Joint Action." In *Embodiment in Evolution and Culture*, ed. Gregor Etzelmüller and Christian Tewes. De Gruyter, 2016.

Gallagher, Shaun, and Dan Zahavi. *The Phenomenological Mind*. 3rd ed. Rout-ledge, 2020.

Gatens, Moira. *Feminism and Philosophy: Perspectives on Difference and Equal-ity*. Indiana University Press, 1991.

Gatens, Moira. "Feminism as 'Password': Re-thinking the 'Possible' with Spi-noza and Deleuze." *Hypatia* 15, no. 2 (2000): 59–75.

Gibson, James J. *The Ecological Approach to Visual Perception*. Houghton Mif-flin, 1979.

Gilbert, Margaret. *On Social Facts*. Princeton University Press, 1992.

Gilligan, Carol. *In a Different Voice: Psychological Theory and Women's Develop-ment*. Harvard University Press, 1982.

Gilligan, Carol. "In a Different Voice: Women's Conceptions of Self and of Morality." *Harvard Educational Review* 47, no. 4 (1977): 481–517.

Glăveanu, Vlad Petre. "Rewriting the Language of Creativity: The Five A's Framework." *Review of General Psychology* 17, no. 1 (2013): 69–81.

Godfrey-Smith, Peter. "On the Status and Explanatory Structure of Developmental Systems Theory." In *Cycles of Contingency: Developmental Systems and Evolution*, ed. Susan Oyama, Paul E. Griffiths, and Russell D. Gray. MIT Press, 2001.

Goldstone, Robert L., and Georg Theiner. "The Multiple, Interacting Levels of Cognitive Systems (MILCS) Perspective on Group Cognition." *Philosophical Psychology* 30, no. 3 (2017): 338–72.

Haken, Hermann. "Introduction to Synergetics." In *Synergetics: Cooperative Phenomena in Multi-Component Systems*, ed. Hermann Haken. Vieweg+Teubner, 1973.

Haken, Hermann, J. A. Scott Kelso, and Herbert Bunz. "A Theoretical Model of Phase Transitions in Human Hand Movements." *Biological Cybernetics* 51, no. 5 (1985): 347–56.

Hanson, Karen. *The Self Imagined: Philosophical Reflections on the Social Character of Psyche* (Routledge, 1988).

Haraway, Donna J. "A Manifesto for Cyborgs." *Socialist Review* 80 (1985): 65–108.

Haraway, Donna J. *Staying with the Trouble: Making Kin in the Chthulucene.* Duke University Press, 2016.

Haraway, Donna J. *When Species Meet.* University of Minnesota Press, 2008.

Haslanger, Sally. "Gender and Race: (What) Are They? (What) Do We Want Them to Be?" *Noûs* 34 (2000): 31–55.

Haugeland, John. *Having Thought: Essays in the Metaphysics of Mind.* Harvard University Press, 1998.

Hausdorff, Jeffrey M., C.-K. Peng, Zvi Ladin, Jane Y. Wei, and Ary L. Goldberger. "Is Walking a Random Walk? Evidence for Long-Range Correlations in Stride Interval of Human Gait." *Journal of Applied Physiology* 78, no. 1 (1995): 349–58.

Heersmink, Richard. "Varieties of the Extended Self." *Consciousness and Cognition* 85 (2020): 103001.

Heft, Harry. "Ecological Psychology and Enaction Theory: Divergent Groundings." *Frontiers in Psychology* 11 (2020): 991.

Heft, Harry. *Ecological Psychology in Context: James Gibson, Roger Barker, and the Legacy of William James's Radical Empiricism.* Psychology Press, 2001.

Heidegger, Martin. *Sein und Zeit*. 17th ed. Max Niemeyer, 1993. Originally published in 1927.

Heinämaa, Sara. "Sex, Gender and Embodiment." In *The Oxford Handbook of Contemporary Phenomenology*, ed. by Dan Zahavi. Oxford University Press, 2012.

Hempel, Carl G., and Paul Oppenheim. "Studies in the Logic of Explanation." *Philosophy of Science* 15, no. 2 (1948): 135–75.

Heras-Escribano, Manuel. *The Philosophy of Affordances*. Palgrave Macmillan, 2019.

Herrnstein, Richard J., and Charles Murray. *The Bell Curve: Intelligence and Class Structure in American Life*. Simon & Schuster, 1994.

Holden, John G., Guy C. Van Orden, and Michael T. Turvey. "Dispersion of Response Times Reveals Cognitive Dynamics." *Psychological Review* 116, no. 2 (2009): 318.

Huebner, Bryce. *Macrocognition: A Theory of Distributed Minds and Collective Intentionality*. Oxford University Press, 2013.

Husserl, Edmund. *Edmund Husserl on the Phenomenology of the Consciousness of Internal Time*. Trans. John Brough. Kluwer, 1991. Originally published as "Edmund Husserls Vorlesungen zur Phänomenologie des Inneren Zeitbewusstseins," in *Jahrbuch für Philosophie und phänomenologische Forschung*, vol. 9, edited by Martin Heidegger. Max Niemeyer, 1928.

Husserl, Edmund. *Ideen zu einer reinen Phänomenologie und phänomenologischen Philosophie*. Edited by Elisabeth Ströker. Felix Meiner, 1992. Originally published in 1913.

Hutchins, Edwin. *Cognition in the Wild*. MIT Press, 1995.

Hutto, Daniel D. *Folk Psychological Narratives: The Sociocultural Basis of Understanding Reasons*. MIT Press, 2012.

Hutto, Daniel D., and Erik Myin. *Evolving Enactivism: Basic Minds Meet Content*. MIT Press, 2017.

Hutto, Daniel D., and Erik Myin. *Radicalizing Enactivism: Basic Minds Without Content*. MIT Press, 2013.

Hyslop, Alec. *Other Minds*. Vol. 246. Springer, 2013.

Ihlen, Espen A. F., and Beatrix Vereijken. "Interaction-Dominant Dynamics in Human Cognition: Beyond $1/f\alpha$ Fluctuation." *Journal of Experimental Psychology: General* 139, no. 3 (2010): 436.

Ingold, Tim. "Anthropological Affordances." In *Affordances in Everyday Life: A Multidisciplinary Collection of Essays*, ed. Zakariah Djebbara. Springer, 2022.

Ingold, Tim. *Making: Anthropology, Archaeology, Art and Architecture.* Routledge, 2013.

Ingold, Tim. "The Textility of Making." *Cambridge Journal of Economics* 34 (2010): 91–102.

Isaac, Alistair M. C. "Modeling Without Representation." *Synthese* 190, no. 16 (2013): 3611–23.

Issartel, Johann, Ludovic Marin, Philippe Gaillot, Thibaut Bardainne, and Marielle Cadopi. "A Practical Guide to Time-Frequency Analysis in the Study of Human Motor Behavior: The Contribution of Wavelet Transform." *Journal of Motor Behavior* 38, no. 2 (2006): 139–59. https://doi.org/10.3200/JMBR.38.2.139-159.

Iverson, Jana M., and Esther Thelen. "Hand, Mouth and Brain. The Dynamic Emergence of Speech and Gesture." *Journal of Consciousness Studies* 6, no. 11–12 (1999): 19–40.

James, William. *The Principles of Psychology.* Henry Holt, 1890. https://psychclassics.yorku.ca/James/Principles/index.htm.

Jayawickreme, Eranda, and Anthony Chemero. "Ecological Moral Realism: An Alternative Theoretical Framework for Studying Moral Psychology." *Review of General Psychology* 12, no. 2 (2008): 118–26.

Juchniewicz, Jay. "The Influence of Physical Movement on the Perception of Musical Performance." *Psychology of Music* 36, no. 4 (2008): 417–27. https://doi.org/10.1177/0305735607086046.

Käufer, Stephan, and Anthony Chemero. *Phenomenology: An Introduction.* 2nd ed. Wiley, 2021.

Kello, Christopher T., Simone Dalla Bella, Butovens Médé, and Ramesh Balasubramaniam. "Hierarchical Temporal Structure in Music, Speech and Animal Vocalizations: Jazz Is Like a Conversation, Humpbacks Sing Like Hermit Thrushes." *Journal of the Royal Society Interface* 14, no. 135 (2017): 20170231.

Kello, Christopher T., Guy C. Van Orden, John G. Holden, and Gregory D. A. Brown. "The Pervasiveness of 1/*f* Scaling in Speech Reflects the Metastable Basis of Cognition." *Cognitive Science* 32, no. 7 (2008): 1217–31.

Kelso, J. A. Scott. "The Haken–Kelso–Bunz (HKB) Model: From Matter to Movement to Mind." *Biological Cybernetics* 115 (2021): 305–22.

Kelso, J. A. Scott. "Phase Transitions and Critical Behavior in Human Bimanual Coordination." *American Journal of Physiology-Regulatory, Integrative and Comparative Physiology* 246, no. 6 (1984): R1000–R1004.

Kelso, J. A. Scott. "Synergies: Atoms of Brain and Behavior." In *Progress in Motor Control: A Multidisciplinary Perspective*, ed. Dagmar Sternad. Springer, 2009.

Kimmel, Michael. "The Micro-Genesis of Interpersonal Synergy." *Ecological Psychology* 33 (2021): 106–45.

Knuuttila, Tarja. "Modelling and Representing: An Artefactual Approach to Model-Based Representation." *Studies in History and Philosophy of Science Part A* 42, no. 2 (2011): 262–71.

Krieger, Michael J. B., Jean-Bernard Billeter, and Laurent Keller. "Ant-Like Task Allocation and Recruitment in Cooperative Robots." *Nature* 406, no. 6799 (2000): 992.

Krueger, Joel. "Direct Social Perception." In *The Oxford Handbook of 4E Cognition*, ed. Albert Newen, Leon de Bruin, and Shaun Gallagher. Oxford University Press, 2018.

Krueger, Joel. "Merleau-Ponty on Shared Emotions and the Joint Ownership Thesis." *Continental Philosophy Review* 46 (2013): 509–31.

Kugler, Peter N., J. A. Scott Kelso, and Michael T. Turvey. "Coordinative Structures as Dissipative Structures I. Theoretical Lines of Convergence." In *Tutorials in Motor Behavior*, ed. George E. Stelmach and Jean Requin. North-Holland, 1980.

Kugler, Peter N., and Michael T. Turvey. *Information, Natural Law, and the Self-Assembly of Rhythmic Movement*. Lawrence Erlbaum, 1987.

Kyselo, Miriam. "The Body Social: An Enactive Approach to the Self." *Frontiers in Psychology* 5 (2014): 986.

Kyselo, Miriam. "The Minimal Self Needs a Social Update." *Philosophical Psychology* 29, no. 7 (2016): 1057–65.

Langland-Hassan, Peter. *Explaining Imagination*. Oxford University Press, 2020.

Latash, Mark L. *Synergy*. Oxford University Press, 2008.

Lee, David N., and E. Aronson. "Visual Proprioceptive Control of Standing in Human Infants." *Perception & Psychophysics* 15 (1974): 529–32.

Lee, David N., and J. R. Lishman. "Visual Proprioceptive Control of Stance." *Journal of Human Movement Studies* 1 (1975): 87–95.

Liao, Shen-Yi, and Vanessa Carbonell. "Materialized Oppression in Medical Tools and Technologies." *American Journal of Bioethics* (2022). https://doi.org/10.1080/15265161.2022.2044543.

Liao, Shen-Yi, and Bryce Huebner. "Oppressive Things." *Philosophy and Phenomenological Research* 103, no. 1 (2021): 92–113.

Loaiza, Juan M. "From Enactive Concern to Care in Social Life: Towards an Enactive Anthropology of Caring." *Adaptive Behavior* 27, no. 1 (2019): 17–30. https://doi.org/10.1177/1059712318800673.

Lobo, Lorena, Manuel Heras-Escribano, and David Travieso. "The History and Philosophy of Ecological Psychology." *Frontiers in Psychology* 9 (2018): 2228.

Locke, John. *An Essay Concerning Human Understanding*. Project Gutenberg, 2004. Originally published in 1690. https://www.gutenberg.org/files /10615/10615-h/10615-h.htm.

Locke, John. *Two Treatises of Government*. Project Gutenberg, 2005. Originally published in 1690. https://www.gutenberg.org/files/7370/7370-h/7370-h .htm.

Ludwig, Kirk. *From Individual to Plural Agency: Collective Action: Volume 1*. Oxford University Press, 2016.

Ludwig, Kirk. "Is Distributed Cognition Group Level Cognition?" *Journal of Social Ontology* 1, no. 2 (2015): 189–224. https://doi.org/10.1515/jso-2015-0001.

Ludwig, Kirk. *From Plural to Institutional Agency: Collective Action II*. Oxford University Press, 2017.

Mach, Ernst. *The Analysis of Sensations and the Relation of the Physical to the Psychical*. Trans. C. M. Williams. Open Court, 1914. Originally published as *Beiträge zur Analyse der Empfindungen* (Gustav Fischer, 1886).

Mackenzie, Catriona, and Natalie Stoljar, eds. *Relational Autonomy: Feminist Perspectives on Autonomy, Agency, and the Social Self*. Oxford University Press, 2000.

Maibom, Heidi L., ed. *Empathy and Morality*. Oxford University Press, 2014.

Maiese, Michelle. *Autonomy, Enactivism, and Mental Disorder: A Philosophical Account*. Routledge, 2022.

Maiese, Michelle. "Embodiment, Sociality, and the Life Shaping Thesis." *Phenomenology and the Cognitive Sciences* 18, no. 2 (2019): 353–74.

Maiese, Michelle. *Embodiment, Emotion, and Cognition*. Palgrave Macmillan, 2011.

Malafouris, Lambros. *How Things Shape the Mind: A Theory of Material Engagement*. MIT Press, 2013.

Marx, Karl. *Theses on Feuerbach*. Trans. Cyril Smith, 2002. Marx-Engels Internet Archive. Originally published in 1845. https://www.marxists.org /archive/marx/works/1845/theses/theses.htm.

Maturana, Humberto R. "Biology of Language: The Epistemology of Reality." In *Psychology and Biology of Language and Thought: Essays in Honor of Eric Lenneberg*, ed. George A. Miller and Elizabeth Lenneberg. Academic Press, 1978.

McDougall, William. *The Group Mind*. Cambridge University Press, 1921.

McGann, Marek. "Enacting a Social Ecology: Radically Embodied Intersubjectivity." *Frontiers in Psychology* 5 (2014): 1321.

McKinney, Joshua, Sune Vork Steffensen, and Anthony Chemero. "Practice, Enactivism, and Ecological Psychology." *Adaptive Behavior* 31, no. 2 (2023): 143–49.

Mead, George Herbert. *Mind, Self, and Society.* Vol. 111. University of Chicago Press, 1934.

Mead, George Herbert. "The Social Settlement: Its Basis and Function." *University of Chicago Record* 12, no. 8 (1907): 108–10.

Menand, Louis. *The Metaphysical Club: A Story of Ideas in America.* Farrar, Straus and Giroux, 2002.

Merleau-Ponty, Maurice. *Phenomenology of Perception.* Trans. Donald A. Landes. Routledge, 2012. Originally published as *Phénoménologie de la perception* (Gallimard, 1945).

Merleau-Ponty, Maurice. *Résumés de Cours: Collège de France, 1952–1960.* Gallimard, 1968.

Merleau-Ponty, Maurice. *The Structure of Behavior.* Trans. Alden L. Fisher. Duquesne University Press, 1963. Originally published as *La structure du comportement* (Presses Universitaires de France, 1942).

Merleau-Ponty, Maurice. *The Visible and the Invisible.* Trans. Alphonso Lingis. Northwestern University Press, 1968.

Morrison, Margaret. *Reconstructing Reality: Models, Mathematics, and Simulations.* Oxford University Press, 2015.

Morrison, Margaret, and Mary S. Morgan. "Models as Mediating Instruments." In *Models as Mediators: Perspectives on Natural and Social Science*, ed. Mary S. Morgan and Margaret Morrison. Cambridge University Press, 1999.

Müller, Robin M. "The Logic of the Chiasm in Merleau-Ponty's Early Philosophy." *Ergo* 4 (2017). https://doi.org/10.3998/ergo.12405314.0004.007.

Nalepka, Patrick, Rachel W. Kallen, Anthony Chemero, Elliot Saltzman, and Michael J. Richardson. "Herd Those Sheep: Emergent Multiagent Coordination and Behavioral-Mode Switching." *Psychological Science* 28, no. 5 (2017): 630–50.

Nalepka, Patrick, Mason Lamb, Rachel W. Kallen, et al. "Human Social Motor Solutions for Human–Machine Interaction in Dynamical Task Contexts." *Proceedings of the National Academy of Sciences* 116, no. 4 (2019): 1437–46.

Nalepka, Patrick, C. Riehm, C. B. Mansour, Anthony Chemero, and Michael J. Richardson. "Investigating Strategy Discovery and Coordination in a

Novel Virtual Sheep Herding Game Among Dyads." In *Proceedings of the 37th Annual Meeting of the Cognitive Science Society* 37 (2015): 1703–8.

Nalepka, Patrick, Paula L. Silva, Rachel W. Kallen, et al. "Task Dynamics Define the Contextual Emergence of Human Corralling Behaviors." *PLoS ONE* 16, no. 11 (2021): e0260046.

Neisser, Ulric. *Cognitive Psychology: Classic Edition.* Psychology Press, 2014. Originally published in 1968.

Neisser, Ulric. "Five Kinds of Self-Knowledge." *Philosophical Psychology* 1, no. 1 (1988): 35–59.

Neisser, Ulric. "The Self Perceived." In *The Perceived Self: Ecological and Interpersonal Sources of Self-Knowledge*, ed. Ulric Neisser. Cambridge University Press, 1993.

Nie, Lin, Dobromir G. Dotov, and Anthony Chemero. "Readiness-to-Hand, Extended Cognition, and Multifactality." In *Proceedings of the Annual Meeting of the Cognitive Science Society* 33, (2011): 1835–40.

Noë, Alva. *The Entanglement: How Art and Philosophy Make Us What We Are.* Princeton University Press, 2023.

Noë, Alva. *Strange Tools: Art and Human Nature.* Hill and Wang, 2015.

O'Regan, J. Kevin, and Alva Noë. "A Sensorimotor Account of Vision and Visual Consciousness." *Behavioral and Brain Sciences* 24, no. 5 (2001): 939–73.

Órnelas, Mark. "Moral Affordances." PhD diss., University of Cincinnati, 2024.

Oyama, Susan, Russell D. Gray, and Paul E. Griffiths, eds. *Cycles of Contingency: Developmental Systems and Evolution.* MIT Press, 2003.

Palermos, S. Orestis. "The Dynamics of Group Cognition." *Minds and Machines* 26, no. 4 (2016): 409–40.

Parr, Adrian, ed. *Deleuze Dictionary Revised Edition.* Edinburgh University Press, 2010.

Paxton, Alexandra, Julia J. C. Blau, and Mikayla L. Weston. "The Case for Intersectionality in Ecological Psychology." https://osf.io/preprints/psyarxiv/jtmea_v1.

Paxton, Alexandra, and Rick Dale. "Argument Disrupts Interpersonal Synchrony." *Quarterly Journal of Experimental Psychology* 66, no. 11 (2013): 2092–2102.

Peck, Zachary. "Towards a Cyborgian Social Ontology of AIPhD diss., University of Cincinnati, 2025.

Peck, Zachary, and Anthony Chemero. "Questioning Two Common Assumptions Concerning Group Agency and Group Cognition." *Proceedings of the 46th Annual Meeting of the Cognitive Science Society*, 46 (2024), 5266–72.

Peng, C.-K., Shlomo Havlin, Jeffrey M. Hausdorff, Joseph E. Mietus, H. Eugene Stanley, and Ary L. Goldberger. "Fractal Mechanisms and Heart Rate Dynamics: Long-Range Correlations and Their Breakdown with Disease." *Journal of Electrocardiology* 28 (1995): 59–65.

Penny, Simon. *Making Sense: Cognition, Computing, Art, and Embodiment.* MIT Press, 2018.

Poldrack, Russell A. "The Physics of Representation." *Synthese* 199, no. 1–2 (2021): 1307–25.

Protevi, John. *Life, War, Earth: Deleuze and the Sciences.* University of Minnesota Press, 2013.

Quintero, Ana María, and Hanne De Jaegher. "Pregnant Agencies: Movement and Participation in Maternal–Fetal Interactions." *Frontiers in Psychology* 11 (2020). https://doi.org/10.3389/fpsyg.2020.01977.

Raja, Vicente. "A Theory of Resonance: Towards an Ecological Cognitive Architecture." *Minds and Machines* 28 (2018): 29–51.

Raja, Vicente, and Michael L. Anderson. "Radical Embodied Cognitive Neuroscience." *Ecological Psychology* 31, no. 3 (2019): 166–81.

Ramseyer, Fabian, and Wolfgang Tschacher. "Synchrony: A Core Concept for a Constructivist Approach to Psychotherapy." *Constructivism in the Human Sciences* 11 (2006): 150–71.

Rawls, John. *A Theory of Justice.* Harvard University Press, 1971.

Reddy, Vasudevi. "On Being the Object of Attention: Implications for Self–Other Consciousness." *Trends in Cognitive Sciences* 7, no. 9 (2003): 397–402.

Reddy, Vasudevi. *How Infants Know Minds.* Harvard University Press, 2008.

Reeve, R., B. Webb, A. Horchler, G. Indiveri, and R. Quinn. "New Technologies for Testing a Model of Cricket Phonotaxis on an Outdoor Robot." *Robotics and Autonomous Systems* 51, no. 1 (2005): 41–54.

Richardson, Michael J., Kerry L. Marsh, Robert W. Isenhower, Justin R. L. Goodman, and R. C. Schmidt. "Rocking Together: Dynamics of Intentional and Unintentional Interpersonal Coordination." *Human Movement Science* 26, no. 6 (2007): 867–91.

Rietveld, Erik, and Julian Kiverstein. "A Rich Landscape of Affordances." *Ecological Psychology* 26, no. 4 (2014): 325–52.

Rigoli, Lily M., Daniel Holman, Michael J. Spivey, and Christopher T. Kello. "Spectral Convergence in Tapping and Physiological Fluctuations: Coupling and Independence of $1/f$ Noise in the Central and Autonomic Nervous Systems." *Frontiers in Human Neuroscience* 8 (2014): 713.

Riley, Michael A., and John G. Holden. "Dynamics of Cognition." *WIREs Cognitive Science* 3 (2012): 593–606.

Riley, Michael A., Michael J. Richardson, Kevin Shockley, and Vanessa C. Ramenzoni. "Interpersonal Synergies." *Frontiers in Psychology* 2 (2011): 38.

Ross, Wendy, Vlad P. Glăveanu, and Anthony Chemero. "The Illusion of Freedom: Towards an Embedded Notion of Constraints." In *Creativity and Constraints*, ed. Catrinel Tromp, Robert J. Sternberg, and Don Ambrose. Brill, 2023.

Ross, Wendy, and Frédéric Vallée-Tourangeau. "Kinenoetic Analysis: Unveiling the Material Traces of Insight." *Methods in Psychology* 5 (2021): 100069.

Sanches de Oliveira, Guilherme. "Radical Artifactualism." *European Journal for Philosophy of Science* 12, no. 2 (2022): 36.

Sanches de Oliveira, Guilherme. "Representationalism Is a Dead End." *Synthese* (2018). https://doi.org/10.1007/s11229-018-01995-9.

Sanches de Oliveira, Guilherme, Vicente Raja, and Anthony Chemero. "Radical Embodied Cognitive Science and 'Real Cognition.'" *Synthese* 198 (2021): 115–36.

Satne, Glenda. "Understanding Others by Doing Things Together: An Enactive Account." *Synthese* 198, suppl. 1 (2021): 507–28.

Schmidt, R. C., Claudia Carello, and Michael T. Turvey. "Phase Transitions and Critical Fluctuations in the Visual Coordination of Rhythmic Movements Between People." *Journal of Experimental Psychology: Human Perception and Performance* 16, no. 2 (1990): 227.

Schmidt, R. C., Lin Nie, Antonio Franco, and Michael J. Richardson. "Bodily Synchronization Underlying Joke Telling." *Frontiers in Human Neuroscience* 8 (2014): 633.

Sebanz, Natalie, and Günther Knoblich. "Progress in Joint-Action Research." *Current Directions in Psychological Science* 30, no. 2 (2021): 138–43.

Segundo-Ortin, Miguel, and Vicente Raja. *Ecological Psychology*. Cambridge University Press, 2024.

Segundo-Ortin, Miguel, and Glenda Satne. "Sharing Attention, Sharing Affordances: From Dyadic Interaction to Collective Information." In *Access and Mediation*, ed. Michael Wehrle, Diego D'Angelo, and Elena Solomonova. De Gruyter, 2022.

Sen, Jobathan, and Darryl McGill. "Fractal Analysis of Heart Rate Variability as a Predictor of Mortality: A Systematic Review and Meta-Analysis." *Chaos: An Interdisciplinary Journal of Nonlinear Science* 28, no. 7 (2018). https://doi.org/10.1063/1.5038818.

Shockley, Kevin, M.-V. Santana, and Carol A. Fowler. "Mutual Interpersonal Postural Constraints Are Involved in Cooperative Conversation." *Journal of Experimental Psychology: Human Perception and Performance* 29 (2003): 326–32.

Silva, Paula, Adam Kiefer, Michael A. Riley, and Anthony Chemero. "Trading Perception and Action for Complex Cognition: Application of Theoretical Principles from Ecological Psychology to the Design of Interventions for Skill Learning." In *Handbook of Embodied Cognition and Sport Psychology*, ed. Massimiliano Cappuccio. MIT Press, 2019.

Smith, Daniel W. *Essays on Deleuze.* Edinburgh University Press, 2012.

Spivey, Michael J. "Cognitive Science Progresses Toward Interactive Frameworks." *Topics in Cognitive Science* 15, no. 2 (2023): 219–54.

Spivey, Michael J. *Who You Are: The Science of Connectedness.* MIT Press, 2020.

Stapleton, Mog. "Enactivism Embraces Ecological Psychology." *Constructivist Foundations* 11 (2016): 325–27.

Steffensen, Sune Vork, and Michael I. Harvey. "Ecological Meaning, Linguistic Meaning, and Interactivity." *Cognitive Semiotics* 11, no. 1 (2018): 1–21.

Steffensen, Sune Vork, Frédéric Vallée-Tourangeau, and Gaëlle Vallée-Tourangeau. "Cognitive Events in a Problem-Solving Task: A Qualitative Method for Investigating Interactivity in the 17 Animals Problem." *Journal of Cognitive Psychology* 28, no. 1 (2016): 79–105.

Stephen, Damian G., and James A. Dixon. "The Self-Organization of Insight: Entropy and Power Laws in Problem Solving." *Journal of Problem Solving* 2, no. 1 (2009): 72–102.

Stephen, Damian G., and James A. Dixon. "Strong Anticipation: Multifractal Cascade Dynamics Modulate Scaling in Synchronization Behaviors." *Chaos, Solitons & Fractals* 44, no. 1–3 (2011): 160–68.

Sutton, John. "Material Agency, Skills and History: Distributed Cognition and the Archaeology of Memory." In *Material Agency: Towards a Non-Anthropocentric Approach*, ed. Carl Knappett and Lambros Malafouris. Springer, 2008.

Swenson, Rod. "Autocatakinetics, Yes—Autopoiesis, No: Steps Towards a Unified Theory of Evolutionary Ordering." *International Journal of General Systems* 21, no. 2 (1992): 207–28.

Swenson, Rod, and Michael T. Turvey. "Thermodynamic Reasons for Perception-Action Cycles." *Ecological Psychology* 3, no. 4 (1991): 317–48.

Theiner, Georg, Colin Allen, and Robert L. Goldstone. "Recognizing Group Cognition." *Cognitive Systems Research* 11, no. 4 (2010): 378–95.

Thompson, Evan, ed. *Between Ourselves: Second-Person Issues in the Study of Consciousness.* Imprint Academic, 2001.

Thompson, Evan. *Mind in Life: Biology, Phenomenology, and the Sciences of Mind.* Harvard University Press, 2007.

Tollefsen, Deborah, and Rick Dale. "Naturalizing Joint Action: A Process-Based Approach." *Philosophical Psychology* 25, no. 3 (2012): 385–407.

Tollefsen, Deborah P., Rick Dale, and Alexandra Paxton. "Alignment, Transactive Memory, and Collective Cognitive Systems." *Review of Philosophy and Psychology* 4 (2013): 49–64.

Thomson, Erik, and Gualtiero Piccinini. "Neural Representations Observed." *Minds and Machines* 28 (2018): 191–235.

Thompson, Evan, and Mog Stapleton. "Making Sense of Sense-Making." *Topoi* (2008). https://doi.org/10.1007/s11245-008-9043-2.

Tronto, Joan C. "Beyond Gender Difference to a Theory of Care." *Signs: Journal of Women in Culture and Society* 12, no. 4 (1987): 644–63.

Turvey, Michael T. *Lectures on Perception: An Ecological Perspective.* Routledge, 2018.

Turvey, Michael T., H. L. Fitch, and B. Tuller. "The Bernstein Perspective: I. The Problems of Degrees of Freedom and Context-Conditioned Variability." In *Human Motor Behavior: An Introduction*, ed. J. A. Scott Kelso. Lawrence Erlbaum, 1982.

Valsiner, Jaan, and René van der Veer. "On the Social Nature of Human Cognition: An Analysis of the Shared Intellectual Roots of George Herbert Mead and Lev Vygotsky." *Journal for the Theory of Social Behaviour* 18, no. 1 (1988): 117–36.

Van Dijk, Ludger, and Erik Rietveld. "Foregrounding Sociomaterial Practice in Our Understanding of Affordances: The Skilled Intentionality Framework." *Frontiers in Psychology* 8 (2017). https://doi.org/10.3389/fpsyg.2016.01969.

van Fraassen, Bas C. *The Empirical Stance.* Yale University Press, 2002.

van Fraassen, Bas C. *The Scientific Image.* Oxford University Press, 1980.

van Fraassen, Bas C. *Scientific Representation: Paradoxes of Perspective.* Oxford University Press, 2008.

van Grunsven, Janna. "Enactivism, Second-Person Engagement and Personal Responsibility." *Phenomenology and the Cognitive Sciences* 17 (2018): 131–56.

van Orden, Guy C., John G. Holden, and Michael T. Turvey. "Human Cognition and 1/f Scaling." *Journal of Experimental Psychology: General* 134 (2005): 117–23.

van Orden, Guy C., John G. Holden, and Michael T. Turvey. "Self-Organization of Cognitive Performance." *Journal of Experimental Psychology: General* 132 (2003): 331–51.

van Rooij, Iris, Raoul M. Bongers, and Pim Haselager. "A Non-Representational Approach to Imagined Action." *Cognitive Science* 26, no. 3 (2002): 345–75.

Varela, Francisco J. *Ethical Know-How: Action, Wisdom, and Cognition.* Stanford University Press, 1999.

Varela, Francisco J. *Principles of Biological Autonomy.* North-Holland, 1979.

Varela, Francisco J., Humberto R. Maturana, and Rodrigo Uribe. "Autopoiesis: The Organization of Living Systems, Its Characterization and a Model." *Cybernetics Forum* 10, no. 2–3 (1981): 7–13.

Varela, Francisco J., Evan Thompson, and Eleanor Rosch. *The Embodied Mind: Cognitive Science and Human Experience.* MIT Press, 1991.

Von Holst, Erich. "Die Koordination der Bewegung bei den Arthropoden in Abhängigkeit von zentralen und peripheren Bedingungen." *Biological Reviews* 10, no. 2 (1935): 234–61.

Vygotsky, Lev S. *Mind in Society: The Development of Higher Psychological Processes.* Harvard University Press, 1978.

Wagenmakers, Eric-Jan, Simon Farrell, and Roger Ratcliff. "Human Cognition and a Pile of Sand: A Discussion on Serial Correlations and Self-Organized Criticality." *Journal of Experimental Psychology: General* 134, no. 1 (2005): 108.

Washburn, Aubrey, Mariana DeMarco, Simon de Vries, et al. "Dancers Entrain More Effectively Than Non-Dancers to Another Actor's Movements." *Frontiers in Human Neuroscience* 8 (2014). https://doi.org/10.3389/fnhum.2014.00800.

Walton, Ashley E., Michael J. Richardson, Peter Langland-Hassan, and Anthony Chemero. "Improvisation and the Self-Organization of Multiple Musical Bodies." *Frontiers in Psychology* 6 (2015): 313.

Walton, Ashley E., Anthony Washburn, Peter Langland-Hassan, Anthony Chemero, Heidi Kloos, and Michael J. Richardson. "Creating Time: Social Collaboration in Music Improvisation." *Topics in Cognitive Science* 10, no. 1 (2018): 95–119.

Weisberg, Michael. *Simulation and Similarity: Using Models to Understand the World.* Oxford University Press, 2012.

West, Bruce J. *Where Medicine Went Wrong: Rediscovering the Path to Complexity.* Vol. 11. World Scientific, 2006.

West, Bruce J., Elvis L. Geneston, and Paolo Grigolini. "Maximizing Information Exchange Between Complex Networks." *Physics Reports* 468 (2008): 1–99.

Withagen, Rob. *Affective Gibsonian Psychology*. Routledge, 2022.

Withagen, Rob. "The Field of Invitations." *Ecological Psychology* (2023): 1–14.

Withagen, Rob, Harjo J. De Poel, Duarte Araújo, and Gert-Jan Pepping. "Affordances Can Invite Behavior: Reconsidering the Relationship Between Affordances and Agency." *New Ideas in Psychology* 30, no. 2 (2012): 250–58.

Withagen, Rob, and John van der Kamp. "An Ecological Approach to Creativity in Making." *New Ideas in Psychology* 49 (2018): 1–6.

Wright, Edgar, dir. *The World's End*. 2013; Working Title Films and Universal Pictures.

Zahavi, Dan. "Empathy, Embodiment and Interpersonal Understanding: From Lipps to Schutz." *Inquiry* 53, no. 3 (2010): 285–306.

Zahavi, Dan. *Self and Other: Exploring Subjectivity, Empathy, and Shame*. Oxford University Press, 2014.

# INDEX

abduction, 170–79
Abney, Drew, 84–85, 166
agency, 43; adaptive, 22; artificial,
    80; autonomous embodied, 103;
    autonomous intelligent, 141;
    autonomous moral, 137–38, 143;
    moral, 138
Addams, Jane, 113
affect, 50, 148
affective framing, 141. *See also*
    Maiese, Michelle.
affordances, 19–20, 26–34, 98, 106–7,
    109, 145, 147; field of, 30–32, 34,
    145; interpersonal, 33; landscape
    of, 30–32, 145; moral, 145
Allan factor analysis, 166–67
Allen, Colin, 94, 97
amelioration, xi
Aristotle, 144
Arnauld, Antoine, 8
Aronson, Eric, 39

artifactualism, 120–26
Artificial Intelligence (AI), 10,
    150–153; and ChatGPT, 5, 150,
    153; and LLMs, 150, 153.
assemblages, 132–33. *See also*
    Deleuze, Gilles.
attractors, 159–60
autopoiesis, 21, 25–28, 134
autonomy, 21, 23, 27, 134–135,
    137–43, 148. *See also* agency;
    relational autonomy

Baggs, Edward, 98
Baier, Annette, xiv, 138
Barandiaran, Xabier, 30
Barge, John, 83
Beer, Randy, 60, 62
being-for-itself, 16, 46–47
being-for-others, 46–47
being-in-the-world. *See Dasein.*
Birhane, Abeba, 1

*Blair Witch Project, The*, 29
blanks, 4–9, 48, 52, 57, 92, 101, 111, 128, 138, 150–53. See also *World's End, The*
Blau, Julia, 139
Bloom, Paul, 148
Boden, Margaret, 106–7
body schema, 15
body-in-an-environment, xii
body-social problem, 48
Bongers, Raoul, 105
brain in the ass hypothesis, 103–4
Brancazio, Nick, 33, 139
Bruno, Giordano, 107
Buhrmann, Thomas, 30
Bunz, Herbert, 64, 158, 174. *See also* Haken-Kelso-Bunz (HKB) model

Candiotti, Laura, 139
Carbonell, Vanessa, 139
Carrello, Claudia, 79
Cash, Mason, 139–40
Chalmers, David, 94, 98–99
chameleon effect, 83.
Chartrand, Tanya, 83.
ChatGPT. *See* Artificial Intelligence (AI) and ChatGPT.
Chomsky, Noam, ix–x, 11, 117
Ciaunica, Anna, 45–47, 142
Clark, Andy, 57, 94, 98–99, 102–3, 105–6
Code, Lorraine, xiv, 137–40
Colombetti, Giovanna, 50
complexity matching, 84–86, 177–80
complex systems, xii–xiii, 133, 170, 181
component-dominant systems, 73, 87, 175–76

computational theory of mind, 10, 56–57, 95–96, 174, 181
conceptual self. *See* self, conceptual.
Confucius, 144
consequentialism, 143, 146, 148
control parameter, 159
coordination: dynamics, 63–65, 67, 81–85, 101, 157, 164–65; in-phase, 68, 76, 81–82; interpersonal, 65, 105; out-of-phase, 64, 68, 81
Cosmelli, Diego, 102–4
coupled oscillators, 63–64, 67–68
coupled oscillatory containment, 80
coupling functions, 61
creativity, xiii, 106–9, 171
critical fluctuations, 68, 76
cross wavelet spectral analysis, 82, 164–66
Cuffari, Elena, 139
Cummins, Fred, 102–4
cyborg, 133–35. *See also* Haraway, Donna.

Dale, Rick, 84–85, 101
Daly, Anya, 139, 184n17
Dasein, 11–13. *See also* Heidegger, Martin.
de Bruin, Leon, 43
de Haan, Sanneke, 43, 45–46, 142
De Jaegher, Hanne, 22, 139, 142
Deduction, 171–73, 178, 180
Deleuze, Gilles, xiv, 107, 127, 130–33, 135, 139, 201n5; and machines, 132–33. *See also* assemblages and larval subjects.
Dennett, Daniel, x, 43–44
deontology, 143, 146, 148
Dereclenne, Emilien, 107

Descartes, René, ix–xi, 6–11, 35, 58, 128–30
detrended fluctuation analysis (DFA), 78, 161–62, 165
Dewey, John, 113, 116
Di Paolo, Ezequiel, 22, 30, 142
dimensional compression, 162–63
Dingenmanse, Mark, 110
distributed cognitive systems, 93–94, 96
Dixon, Jay, 105
Dotov, Dobromir, 76, 161
Dutt, Anjali, 147

ecological psychology, 19–20
ecological self. *See* self, ecological.
ecological-enactive approach, 32, 34, 66, 139, 141, 144, 199
*Einfühlung*, 49
empathy, 49–51, 148; cognitive, 50, 148; sensorimotor, 50–51, 82, 86, 91, 141, 148
enactivism, 21–23
ethics of care, 146–48
extended self. *See* self, extended.
Exteroception, 36, 38

Fanon, Frantz, 33–34
Farrell, Simon, 174
feminist political theory. *See* political theory, feminist
field of affordances. *See* affordances, field of
field of view, 36–39
fix cycle, 93, 96–97
flat ontology of becoming, 131–32. *See also* Deleuze, Gilles
Fodor, Jerry, 56–57, 170, 174, 179

form of life, 30–32, 34
fractals, 159, 169, 176; and multifractality, 176–80
Frankfurt, Harry, 171
Fusaroli, Ricardo, 84

Gallagher, Shaun, 29–30, 42, 44–45, 51, 86
Gatens, Moira, xiv, 130, 135–37, 139–40, 201n5
Gibson archives, 19
Gibson, James J., 18–20, 36, 38, 40, 58–59; and Gibsonians, 25, 29–30, 109
Gilligan, Carol, xiv, 146–47
Glăveanu, Vladimir, 109
Goldstone, Rob, 94, 98
grip, 31
group cognition, xiii, 92–93, 96–101, 195
Guattari, Félix, 130, 132

Haken, Hermann, 63, 158, 174. *See also* Haken-Kelso-Bunz (HKB) model
Haken-Kelso-Bunz (HKB) model, 64, 67–68, 76, 79
Haraway, Donna, xiv, 133–35, 139
Haselager, Pim, 105
Haslanger, Sally, xi
Haugeland, John, 170, 174, 179; and giving a damn, 153
Heft, Harry, 30
Hegel, G.W.F., 116
Heidegger, Martin, 11–13, 18, 45, 58; and ready-to-hand, 76–78; and unready-to-hand, 76–78
Heinämaa, Sara, 139

Holden, Jay, 172
Holman, Daniel, 177
Holmes, Oliver Wendell, 112
Huebner, Bryce, 139
Hurst exponent, 161–162
Husserl, Edmund, xii, 17, 23
Hutchins, Edwin, 93–95, 97
Hutto, Daniel, 44, 190n5
Huygens, Christiaan, 63, 68
hylomorphic model, 107–8

Ihlen, Espen, 176
*Ineinander*, xii, 17, 24, 98, 184n17
Ingold, Tim, 107–9, 131–32
idealism, 24–25
imagination, 7, 30, 106–9
induction, 171–72, 178, 180
informational encapsulation,
    170–171
inner speech, xiii–xiv, 111, 117–18,
    120, 122–26. *See also* self, private
intentional stance, x
interaction theory, 51, 86. *See also*
    Gallagher, Shaun
interaction-dominant systems,
    70–73, 75–78, 87, 159, 161–62,
    169–170, 173, 175, 177, 179–80
intercorporeality, 16
interpersonal affordances.
    *See* affordances, interpersonal
interpersonal self. *See* self,
    interpersonal
invitations, 27, 29–30, 32–34,
    145–46; moral, 145, 147

James, William, 112, 116, 125
Jayawickreme, Eranda, 144, 147
Jefferson, Thomas, 9

Kello, Christopher, 84–85, 177
Kelso, Scott, 59–60, 64, 67, 158,
    174. *See also* Haken-Kelso-Bunz
    (HKB) model
Khepera robot, 121
kinenoesis, 109
kinesthesis. *See* proprioception
Kiverstein, Julian, 30–32
Knoblich, Günther, 100
Kohlberg, Lawrence, 146
Kohlfeldt, Danielle, 147
Krueger, Joel, 49, 184n17
Kugler, Peter, 59–60
Kyselo, Miriam, 45, 47–48, 51, 142

Laboratory School, the, 113
landscape of affordances.
    *See* affordances, landscape of
Langland-Hassan, Peter, 106–7
larval subjects, 133. *See also* Deleuze,
    Gilles
Lee, David, 39
Liao, Sam (Shen-Yi), 139
liberal political theory. *See* political
    theory, liberal.
Lishman, Roly, 39
lived body, 14–16, 21–22, 50, 78.
    *See also* Merleau-Ponty, Maurice
LLMs. *See* Artificial Intelligence
    (AI) and LLMs
Loaiza, Juan, 139, 147
Locke, John, xiv, 8–11, 35, 127–29, 135
long memory, 72–73, 169
Ludwig, Kirk, 96–99

Mach, Ernst, 37
Mackenzie, Catriona, 140
magnet effect, 63, 157–58

Maiese, Michelle, 139, 141–42
maintenance effect, 63, 158
Malafouris, Lambros, 108, 131–32
Marx, Karl, 89, 116, 149
material engagement theory,
    108–109. *See also* Malafouris,
    Lambros
McDougall, William, 93
Mead, George Herbert, xiii, 113–17,
    122, 125, 200n8
memory. *See* long memory;
    transactive memory systems
mental state, x, 148
Merleau-Ponty, Maurice, xii, 11,
    14–19, 21, 24, 45, 58, 98, 184n16,
    184n17
Metaphysical Club, the,
    112, 199n1
methodological solipsism, 57
minimal self. *See* self, minimal
models. *See* philosophy of
    science and models. *See also*
    artifactualism.
modern theory of the mind, ix–xii,
    xiv, 6–11, 18, 24, 35, 42, 48, 52,
    56–58, 62, 87, 97, 100, 102, 110,
    128–29, 130–32, 135–36
moral invitations. *See* invitations,
    moral
moral maturity scales. *See* Kohlberg,
    Lawrence
Müller, Robin, 184n17
Myin, Erik, 185n8, 190n5

Nalepka, Patrick, 79, 162, 164
narrative self. *See* self, narrative.
Neisser, Ulric, 40–42, 45–46
Nie, Lin, 76

Noë, Alva, 185n8, 186n12
nonlinear dynamical systems theory,
    xii, 39, 57, 91, 108, 158

1/f noise. *See* pink noise
optical contraction, 39
optical expansion, 39
Órnelas, Mark, 147

Panpsychism, 133
Parks, Rosa, 34
Parr, Adrian, 130
participatory sensemaking, 22–24,
    47–48, 134, 147–48, 186n12
Paxton, Alexandra, 84–85, 101, 139
Peirce, Charles Saunders, 112,
    169–173, 175, 178, 180
perception-action cycle, 19, 119–120,
    123
phenomenology, xii, 11, 14, 16, 19–22,
    42, 45, 48, 52, 58, 60, 62, 76, 131, 170
philosophy of science, 120, 123;
    and models, 120–122. *See also*
    artifactualism
pink noise, 77–78, 159, 161–62,
    169–170, 173–78
political theory: feminist, xi,
    127–128, 137; liberal, 10, 127
practical wisdom, 145
pragmaticism, 112
pragmatism, 112–16
principle component analysis (PCA),
    163–64
private self. *See* self, private
problem of other minds, xi–x, 3–6, 8,
    11, 14, 17–18, 48
proprioception, 36, 38
Protevi, John, 130, 133

Rączaszek-Leonardi, Joanna, 84
Radcliff, Roger, 174
radical embodied cognitive science,
    xii, 29–30, 57–58, 62, 71, 87, 99,
    131–32
Ramsmeyer, Fabian, 83
Rawls, John, 135–37; and the
    original position, 136; and the
    veil of ignorance, 136–37
realism, 24–25
Reddy, Vasudevi, 46
relational autonomy, 140–43
relative phase, 158, 160, 165
representation hunger problem, 92
Richardson, Michael, 69, 79, 81, 83
Rietveld, Erik, 30, 32, 145
Rigoli, Lillian, 177
Rosch, Eleanor, 21, 24–25, 29
Ross, Wendy, 109
Rousseau, Jean-Jacques, 135–36
Royce, Josiah, 116

Sartre, Jean-Paul, 46–47
Schmidt, Richard, 79, 164
Sebanz, Natalie, 100
second persons, 138, 140, 144, 146,
    148
self, 9–10; conceptual, 41; ecological,
    44–46, 133; extended, 42;
    interpersonal, 40–41, 45–46,
    51; minimal, xii, 42–48, 51–52,
    92, 98–99, 133, 142; narrative,
    42–43, 45; private, 42, 45
self-organized criticality, 76
sensorimotor abilities, 26–27
sensorimotor empathy. See empathy,
    sensorimotor
settlement houses, 113

settlements, 114
Sherrington, Charles, 36
Shockley, Kevin, 83
Simondon, Gilbert, 107
simulations, 48–51, 66, 86, 120, 148,
    175–76
Skilled Intentionality Framework,
    27, 30–31, 33
Smith, Daniel, 130
Spivey, Michael, 110, 177, 180
Stapleton, Mog, 29
Stephen, Damian, 105
Stoljar, Natalie, 140
Swenson, Rod, 25
sympoiesis, 134
synapse, 36
synergy, xii–xiii, 69–72, 77–83,
    162–64

Theiner, Georg, 94, 96–98
thinging, 109, 117
Thompson, Evan, 21, 24–25, 104
Titchener, E. B., 49
Tollefsen, Deborah, 101
Toribio, Josefa, 105–6
transactive memory systems, 95,
    97, 101
Tsacher, Wolfgang, 83
Turvey, Michael, 59–60, 79, 170,
    172–74, 179
Tylén, Kristian, 84

Vallée-Tourangeau, Frédéric, 109
van der Kamp, John, 109
van Grunesven, Janna, 139, 147
van Orden, Guy, 172–78
van Rooij, Iris, 105
Vandermeer, Jeff, 53

Varela, Francisco, 21, 24-25, 29, 129
variability, 67, 72, 161, 164. *See also*
    pink noise
Vereiken, Beatrix, 176
virtue ethics, 144-45, 147-48
von Holst, Erich, 63-64, 157
Vygotsky, Lev, xiii, 115-17, 122

Wagenmakers, E. J., 174-78
Walton, Ashley, 81-82, 101

Wegner, Daniel, 94-95
West, Bruce, 84
Weston, Mikayla, 139
Withagen, Rob, 109
Wittgenstein, Ludwig
*World's End, The*, xi, 3, 52, 56, 92, 111
Wundt, Wilhelm, 49

Zahavi, Dan, 43, 45-46
zone of proximal development, 115

GPSR Authorized Representative: Easy Access System Europe, Mustamäe tee 50, 10621 Tallinn, Estonia, gpsr.requests@easproject.com

www.ingramcontent.com/pod-product-compliance
Lightning Source LLC
Chambersburg PA
CBHW032128020426
42334CB00016B/1082